AF847176

Towards Sustainable Urban Transportation

Linas Kliučininkas

Towards Sustainable Urban Transportation

Environmental Dimension

PETER LANG
Frankfurt am Main · Berlin · Bern · Bruxelles · New York · Oxford · Wien

Bibliographic Information published by the Deutsche Nationalbibliothek
The Deutsche Nationalbibliothek lists this publication in the Deutsche Nationalbibliografie; detailed bibliographic data is available in the internet at http://dnb.d-nb.de.

Layout: www.kumpernatz-bromann.de

ISBN 978-3-631-62367-1

© Peter Lang GmbH
Internationaler Verlag der Wissenschaften
Frankfurt am Main 2012
All rights reserved.

All parts of this publication are protected by copyright. Any utilisation outside the strict limits of the copyright law, without the permission of the publisher, is forbidden and liable to prosecution. This applies in particular to reproductions, translations, microfilming, and storage and processing in electronic retrieval systems.

www.peterlang.de

To my parents and family

Contents

Preface

The current research publications in the field of urban transportation show a rapid development towards Sustainable Transportation Systems (STS). The assessment of environmental quality is one of the issues of great concern in STS development. Today, environmental threats posed by transportation are well enough known to the general public. However, principles and procedures on how to evaluate the influence of urban transportation upon human beings and ecosystems, and how to prioritise the necessary measures, are still vague. Practical knowledge of the environmental impacts of urban transportation therefore needs to be widened and improved in order to properly inform decision-making and planning.

This monograph provides systematic, step-by-step theoretical knowledge, as well as case studies obtained via practical data collection, validation, aggregation, storage and finally reporting of the aggregated data for the purpose of management. These procedures go hand-in-hand with computational methods and modelling.

The author hopes this monograph will be of interest to professionals and academics working in the fields of environmental protection and urban traffic planning.

I am grateful to friends and colleagues from the Baltic University Programme (BUP) network for their valuable support, insights and comments during the writing of this monograph.

Linas Kliučininkas
Kaunas
November 2011

I. The challenges of urban transportation

1.1 Trends and features

In all modern societies, for a number of reasons, people tend to increase their mobility. The functioning of a modern society puts heavy demands on the ability of individuals to be mobile. Economic, political and social factors affect both the total volume of passenger transportation and its different modes. The worldwide increase in urban mobility since 1960 has been the direct result of increased affluence and the consequent improved access to private motor vehicles, as well as population growth. Cameron and his team (2003, 2004) investigated the cause of increased urban traffic in world cities between 1960 and 1990, and concluded that population growth, urban sprawl, increased car ownership, and decreased vehicle occupancy are the key factors causing the steep rise in vehicle kilometres (vkm).

In a broader sense, the increase in mobility can be illustrated by five main tendencies:

- longer distances covered in each trip;
- rapid growth of travel by car;
- increased importance of leisure-time activities;
- spatially more complex travel patterns;
- diffusion of mobility from a few groups to the broad majority of the population.

Mobility development in Sweden may be traced back to the beginning of the last century. Around 1900, the average distance travelled per person per day was roughly one kilometre. In the 1950s, mobility grew fast and increased to an average of more than ten kilometres a day. At the end of the last century, in the population as a whole it amounted to 35 kilometres per individual. Therefore over the course of the century people's ability to cover distances and interact with other places has increased by about 35 times (Tengstrom, Thynell, 1997).

Cities and towns act as engines for progress, often driving much of our cultural, intellectual, educational and technological achievements and innovations. However, today's trend of urban development results in increased consumption of energy, resources, transport and land, thereby raising greenhouse gas emissions and air and noise pollution to levels that often exceed the legal or recommended human safety limits. This development, of course, has affected the lifestyles and daily activity patterns of most people, as well as the structure of cities and land-use in general. However, the ability to be mobile is not evenly distributed over the population. We can also observe that, in recent years, mobility has not grown as fast as has previously been the case.

Urban freight transport has to respond to changes in industry structure: in developed countries, it has had to adapt to high value-added production. This means increased value added per unit of weight, and consequently greater demand for more efficient transport. In addition, improved production technologies and management, as well as shorter production development cycles, tend to minimize inventories. Advancements in production and consumption have required smaller shipments and more frequent freight transport. The increased competition brought about by deregulation has stimulated higher quality services responding to these needs. Consequently the actual unit load carried per vehicle and trip has decreased (Nakamura et al., 2004).

However, mobility ought to be measured not only in terms of daily distance covered but also taking into account trip-frequency and the amount of time spent on travel. Daily trip-frequency is a crude measure of social interaction; that is, the number of out-of-home activities of a population. Travel-time per day reflects the role of transport in the daily time-budget; that is, how much time we can devote to travel considering everything else that has to be done.

Experience of development in most European countries suggests that mobility considered as trip-frequency does not change significantly in the long term. This means that people do not travel more and more often: they travel greater and greater distances. A similar observation applies to the travel time budget. People do not devote more time to travel: they travel faster. If people have the opportunity to choose, they generally use the fastest mode of transportation.

For a long period, commuting between home and work was the main purpose of travel and structured traffic-flows into regular patterns in time and space. But, at least in the more wealthy European states, during recent decades leisure trips have shown the biggest increase (defined as visiting friends and relatives, or other social/recreational activities). These "free time" trips now account for between one third and one half of total trips and daily distance covered. So we can see that change in the location of home and work, or the *suburbanization* of city life, represents only a minor part of the evolving mobility pattern. Another aspect of developments in mobility is an increased geographical distribution of activities, in addition to the growing complexity of spatial relations. After a first wave of suburbanisation – spatially concentrated, close to old city centres and well served by public transportation – new patterns have emerged. A second wave has meant an increased attraction for low-density areas – rural places, small towns, outer suburbs – for housing developers. At the same time, there has been a suburbanisation of jobs, shops, and healthcare and leisure services. Many of the new places are not easily served by public transport, and this has engendered new and complex spatial relationships with regard to travel. Traffic within and especially between suburbs ("tangential" trips) has increased, while the old patterns of regular trips from suburbs to city centre ("radial" trips) has stagnated. *Conurbations* are also attracting commuters from increasingly large catchment areas.

Suburbanisation in the US was proceeding rapidly, with more workplaces being progressively relocated to the suburbs of large cities. This resulted in a phenomenon called suburban gridlock, in which traffic congestion occurred not only in the city centres but also in the suburbs. In Europe, where cities are frequently very old, there are many cases where urban areas are very compact. Since it was difficult to construct new trunk roads in such localities, urban traffic policy with regard to motorisation was focused on improving pedestrian spaces and public transport facilities: for example by the construction of transit malls and LRT (Light Rail Transit) lines. Although underground railways were also constructed, most cities opted to continue with the operation of trams, and increased their efforts to make these more convenient and agreeable.

One final significant feature of mobility development is the spread of mobile life-styles from few to many segments of the population, and to different social groups and households. This includes aspects of age and gender. The mobility of the elderly has increased thanks to the simple fact that the first generation of car owners is now reaching old-age. The entrance of women into the labour market during recent decades has of course increased demand for travel among women. But the most important factor by far is that, in the wealthiest countries over the last fifty years, almost 80 per cent of all households have been able to buy a car.

Experiences from most countries show that by far the most important factor in mobility level is car ownership. In countries like Denmark, Germany, Finland and Sweden, the car accounts for 70 to 80 per cent of the total daily distances of passenger travel. Car ownership is largely influenced by household income; other influences include, of course, the cost of buying and then using and maintaining the car. Differing levels of political control, taxes on fuel and vehicles, and charges such as import duties influence demand in different countries.

Is there a saturation level in car ownership? It is hard to give a straightforward answer to this question. In Sweden, there is on average one car per household, which may seem a plausible point of balance. However, in the United States this figure had already been reached in the 1950s. Today there are almost 0.7 cars per person in the US, which means that there is more than one car per licence-holder. But economic factors – such as the price of petrol, the structure of cities and value systems, and general culture – differ. In Europe, there are clear signs that the car is regarded as a source of individual mobility, rather than a commodity related to the household in general. This individualisation of car ownership and use is certainly one possible impetus for further mobility expansion in the wealthier nations.

Regions with high densities of population and settlements tend to have fewer cars per capita and thus lower mobility levels than regions that are sparsely populated and with dispersed settlement patterns. At the same time, the tendency for cities and activities to spread out over large areas is a result of the increased mobility of the population.

Another important factor – at least in densely populated areas – is the standard of public transportation. Areas served by public transportation of high quality – high frequency, high speed, well-integrated routes and networks that are secure, comfortable and reasonably priced – have proportionally lower car ownership levels. If public transportation cannot live up to such relatively high standards it runs the risk of entering a vicious circle. More people turn to the car, leaving public transport to a diminishing group of people who, for various reasons, cannot use other modes: these are the so-called "captive riders".

1.2 Towards sustainable urban transportation

Traditionally in transport policy, three main types of policy instruments have been used to implement concrete measures:

- technical development;
- urban planning;
- economic incentives.

However, it is also important to consider that traditional policy instruments are insufficient to realise the goals of sustainable mobility. Other political measures include legal framework, making information available to the general public, and education. At the same time we should consider that there are a lot of powerful actors – for example oil companies, among others – who are opposed to any significant move away from the present fossil-fuelled car, be it through the introduction of biofuels and/or a change of vehicle concept.

Technical improvements can be discussed on two different levels: long-term technical potential and short-term possibilities. The potential for passenger transport is generally greater than for cargo transport. Improved transport technologies are challenges for car manufacturers. Current technical solutions are geared towards decreasing specific energy use or specific emissions (per vehicle kilometre).

The long-term technical potential for reducing energy use is related to reducing greenhouse emissions.

The internal combustion engine is a mature technology that has undergone a massive amount of technological development, and continues to receive enormous research and development support. It is therefore likely to sustain its dominance of the automobile and lorry market in at least the medium term.

The main attraction of electric vehicles is the complete absence of tail-pipe emissions, and consequently the promise of improved urban air quality. A further advantage is that using electricity to power vehicles affords much greater primary fuel flexibility: this reduces dependence on petroleum products, but importantly may also ultimately enable vehicles to be powered by renewable energies such as solar, wind or hydroelectric power. Consequently, electric vehicles have the potential to be essentially pollution-free on both a well-to-tank and tank-to-wheels basis.

A hybrid vehicle, which takes peak power from the buffer store and requires about five times less power than a conventional car, is another branch of development. One important task is to reduce so-called cold-start emissions, which represent a considerable proportion of emissions from a car equipped with a catalytic converter. Another problem is the reliability of the entire emission-control system during a car's whole life-span.

Technical improvements are necessary but not enough. There are significant correlations between town planning and urban transport, and energy-use and pollution emissions. But while relationships between the availability of urban transport and energy-use and pollution emissions are approximately proportional, the correlation between town planning and urban transport is more complicated. In general, the internal relationships between housing, employment and shopping are important. The longer the distances between these functions, the higher the energy consumption. This is caused by two factors: increased transport work and more journeys by car, both of which increase consumption per mile.

Urban transport is reduced when these relationships are balanced within each urban society and the functions are distributed across the region, whilst concentrating them in regional centres results in increased traffic. Urban transport seems to be minimized in towns and cities of between 200,000 and 300,000 inhabitants with a high number of cars. Households with cars travel twice as much as households without.

Migration from rural areas to cities, further industrialisation, continuously growing consumption, service-sector growth, developments in building technology, plus the explosive increase in private transport: all these factors place demands on cities that cannot be immediately met within the existing frame of operation. Solutions have been sought in the heavy-handed reconstruction or total replacement of town centres. Old industry, workshops and housing have had to yield to the expansion of city occupations. This has altered the basis for urban planning. Improving existing central areas has been made possible through suitable conservation and housing-oriented renovation, as well as by developing necessary public and private services, recreational open spaces and traffic improvement.

At the same time, access to city centres has been increased by the development of extensive traffic systems, including urban motorways linking the new central business districts to existing urban areas as well as to new suburban residential and industrial areas.

There are energy savings to be found by increasing density in built-up areas. It has been shown that, in a town with 30 inhabitants per hectare, the use of energy for transportation is 20 per cent lower than in a town with half that density. It follows that the urban-growth problem should be partly solved by the regeneration of debilitated urban areas and a more intensive use of the existing urban structure.

This has made it possible to keep undeveloped areas close to the city free from urban development, and for them to serve instead as ecological zones contributing toward the city's oxygen renewal, the maintenance of the ecological cycle and the recreation of its citizens.

It has also been shown that the availability of transport and energy consumption increase significantly when the distance between housing and the city centre rises. The energy consumption of households 40 kilometres from the city centre is close to twice as much as that for households located in the city, if local shopping and service facilities are poor; it is 30 per cent higher if there are good local services.

In most countries, a number of different taxes and charges are in place in the road transportation sector. These charges may be interpreted as economic policy instruments, even though many of them have been introduced for other purposes, such as raising revenues for the government. The most common traditional economic policy instruments are the following:

- fuel taxes,
- mileage taxes,
- annual vehicle taxes,
- parking fees,
- subsidies for scrapping old cars.

Making existing cities and new urban development more ecologically based and liveable is an urgent priority in the global push for sustainability. Kenworthy (2006) discusses ten critical responses to this issue and summarises them in a simple conceptual model that places the nexus between transport and urban form at the heart of developing an eco-city. This involves compact, mixed-use urban forms, well-defined higher-density and human-oriented centres, priority given to the development of superior public transport systems and conditions for non-motorised modes, along with minimal road capacity increases, and the protection of the city's natural areas and food producing capacity. These factors form the framework in which everything else is embedded and must operate, and if they are not addressed only marginal changes in urban sustainability can be made. Within this framework, environmental technologies need to be extensively applied. Economic growth needs to emphasise creativity and innovation and to strengthen the environmental, social and cultural amenities of the city. Public areas throughout the city need to be well-designed and maintained, and sustainable urban design principles need to be applied in all urban development. All these dimensions need to operate within two key processes involving vision-oriented and reformist thinking and a strong, community-oriented, democratic sustainability framework for decision-making.

Urban form and spatial development have major consequences for sustainable development, encompassing not only environmental issues but also social and economic aspects. The concept of compact city development aims to optimize energy use, promote renewable energy sources, provide integrated transport networks with a focus on enhancing public transport and cycle routes, change the culture of energy and resource consumption and increase social inclusion (Jenks and Jones, 2010).

Most of the National Strategic Reference Frameworks submitted by the member states of the European Union include sustainable urban transport as an area for action. Extension, rehabilitation and upgrading of clean urban public transport such as trolley buses, trams, metros and suburban rail as well as other sustainable urban transport projects should continue to be promoted and supported by the EU. Pollutant emissions from vehicles have been successfully reduced through a gradual tightening of the EURO emissions standards (Green Paper, 2007). It has become clear during the consultations that in order to create a new "urban mobility culture" in Europe there is a need to set up partnerships. New planning methods and tools can also play a major role in this new urban mobility culture. Education, training and raising awareness have an important role to play.

References

Cameron, I., Kenworthy, J.R. and Lyons, T.J. (2003). Understanding and predicting private motorised urban mobility. *Transportation Research Part D* 8, pp. 267-283.

Cameron, I., Lyons, T.J. and Kenworthy, J.R. (2004). Trends in vehicle kilometres of travel in world cities, 1960-1990: Underlying drivers and policy responses. *Transport Policy* 11, pp. 287-298.

Green Paper (2007). Towards a new culture for urban mobility. Brussels, COM (2007) 551. p. 23.

Jenks, M. and Jones, C. (eds.) (2010). *Dimensions of the sustainable city 2*. London: Springer.

Kenworthy, J. (2006). The Eco-City: Ten Key Transport and Planning Dimensions for Sustainable City Development. *Environment and Urbanization*. Special Issue, 67-85, April.

Nakamura, H., Hayashi, Y. and May, A.D. (eds.) (2004). Urban Transport and the Environment. Elsevier, pp. 60-74.

Tengstrom, E. and Thynell, M. (eds.) (1997). Towards Sustainable Mobility: Transporting people and goods in the Baltic Region. The Baltic University Programme, Uppsala University, p. 5-7.

II. Environmental impacts of road transportation

2.1 Energy consumption

The transport sector is the fastest growing energy-consuming sector in the OECD, and is the most problematic for environmental policy. A rough estimation of energy use in OECD countries might include three constituents: one third of energy is used for residential purposes, one third is used for commercial-industrial purposes, and one third of energy is used for transportation. The fact is that a high proportion of residents of OECD countries prefer to travel by automobile, and most manufacturers find it more convenient to ship their products by lorry. Getting people to switch to public transport is difficult and of course requires substantial investment in infrastructure in order to make alternatives to private travel available in the first place. The effectiveness of fuel taxes is unclear, since it is questionable how much consumers take fuel costs into consideration either when buying a car or when deciding how much to use it. An alternative to a fuel tax is an excise tax on new cars, differentiated by the fuel efficiency of the car, which, though not a popular measure, is thought by some experts to be more effective than fuel taxes in changing consumer behaviour. Also being implemented or under consideration in some countries are road taxes, increased parking fees and restrictions on employment-related incentives such as company cars. As noted, there is also a continuing discussion on fuel economy standards imposed on manufacturers (similar to the Corporate Average Fuel Economy – CAFE standards in the US). As with energy policy, however, individual states have very different transport systems and have correspondingly different ideas about what constitute (politically and economically) feasible policies and measures. The best that can be expected in this sector in the short term is that other problems, such as congestion and air pollution, will prompt the adoption of policies which will, as a side benefit, limit growth in CO_2 emissions.

All OECD countries have, either specifically or as part of their broader climate change programmes, adopted measures intended to encourage greater use of public transport. However, no country has succeeded in increasing the use of public transport relative to private vehicle use. Germany has done the most to make public transport an attractive and viable alternative. Denmark has done the most to discourage road travel through taxes (while at the same time continuing to invest in road infrastructure). Portugal, Spain and Greece are actually supporting the massive expansion of road infrastructure, while the governments of Finland, Sweden and Norway, because of their size and geography, are, by default, encouraging road travel by not investing adequately in alternative modes. Some governments are facing incidents of civil disobedience aimed at stopping road construction and expansion projects (Anderson, D., 1997).

Fig. 1: Newman and Kenworthy's study demonstrating a relationship between overall urban density and transport energy use in world cities

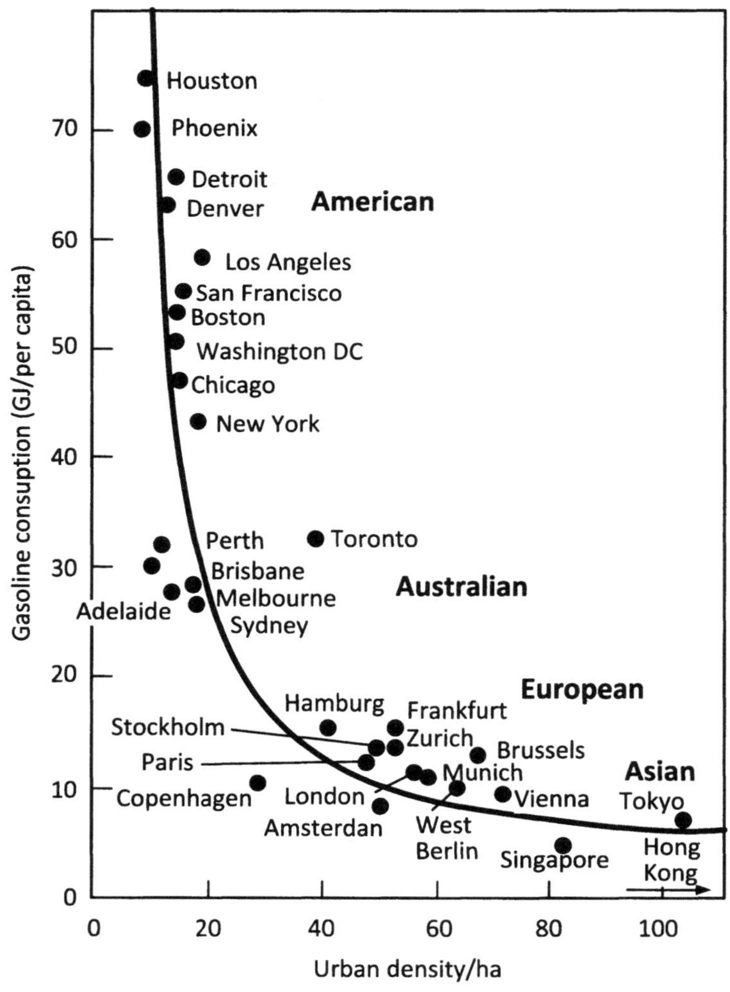

Fig. 2: Revised graph illustrates Newman and Kenworthy's relationship between overall urban density and transport energy use in world cities

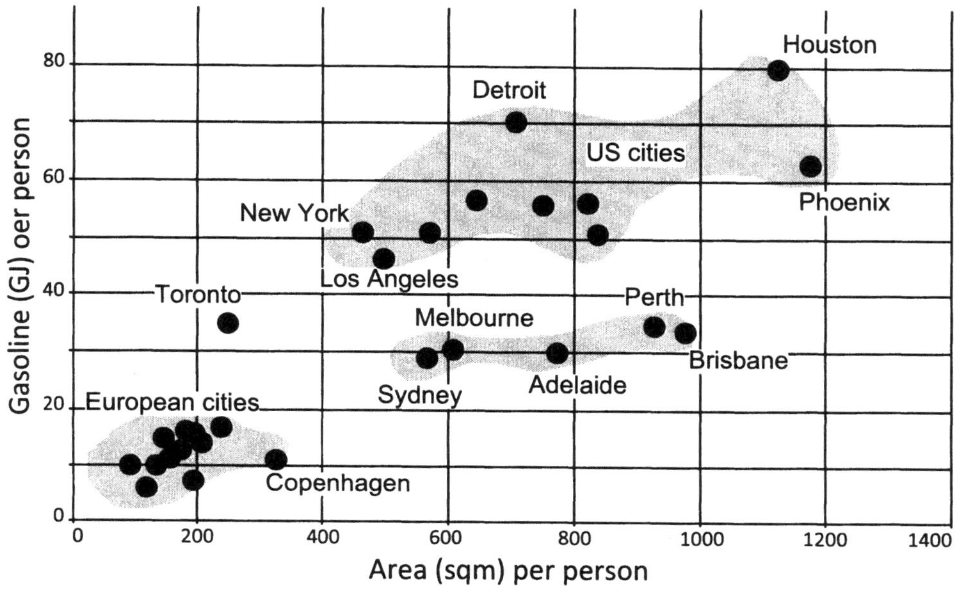

Newman and Kenworthy's study is often referred to when demonstrating a relationship between overall urban density and transport energy use (Newman, P., Kenworthy, J.R., 1998). Gasoline consumption per capita in ten large United States cities varies by up to 40 per cent, primarily because of land-use and transportation planning factors, rather than price or income variations. The same patterns, though more extreme, appear in a global sample of 32 cities (see Fig. 1). Here, average petrol consumption in US cities was nearly twice as high as in Australian cities, four times higher than in European cities and ten times higher than in Asian cities. Variations in petrol price, income, and vehicle efficiency can explain only half of these differences. The authors suggested that physical planning policies, particularly reurbanisation, and a reorientation of transportation priorities as a means of reducing petrol consumption, may also play a role.

Newman and Kenworthy's statistical methods have been criticized by a number of authors. The essential problem is examining the relationship between "motor-spirit use per person" and "persons per hectare". Since population represents the denominator in one term and the numerator in the other, a spurious correlation emerges. Reanalysis of the primary data comparing petrol use per person to area per person shows a more complex relationship. The revised graph shows this (see Fig. 2).

Transport energy amounts to some 20 per cent of total energy use in Europe. More than 90 per cent of it is based on petroleum fuels. About two-thirds of transport energy is used for passenger transport. The transport share of total energy use, as well as the per capita transport energy use, is larger in Western European countries than in most Central and Eastern European countries.

While the increase in total energy use was very moderate in industrialised countries during the 1970s and 1980s, energy use for transport has increased significantly. For example, energy use in the EU doubled between 1970 and 1990, whereas energy use in other sectors showed only a minor increase.

The worldwide increase in urban mobility since 1960 has been the direct result of increased affluence and the consequent greater access to private motor vehicles, as well as population growth.

Technical developments have reduced specific energy use (MJ/vehicle km) during the period, but this reduction has been counteracted by the very large increase in transport volume. For example, passenger transport in Western Europe almost doubled between 1970 and 1990 whereas the transport of goods increased by almost 70 per cent. A large fraction of the potential energy efficiency improvements for vehicles has also been lost due to increases in vehicle weight and power.

There are also additional factors that have led to increased energy consumption. Firstly, transport has tended to use less energy-efficient modes of transport, such as road transport in preference to rail, or air in preference to road. Secondly, vehicles are used less efficiently, since occupancy and load-factors of vehicles are being reduced. Finally, driving conditions have worsened: in particular, roads are increasingly crowded, and this has also contributed to increasing energy use.

Higher average speeds on highways are another factor. Efficient energy use is heavily dependent on vehicle speed. Energy use is lowest at average speeds of approximately 60-80 km/h. At higher speeds energy increases rapidly. This is mostly due to increased air-drag.

In cities it is the other way round. Here the speed is often much lower than optimal. Energy use at low average speeds in urban traffic is inefficient, and consumption is higher. There is greater engine friction, more retardation and more idling than in more fluent traffic conditions (Tengstrom, E., Thynell, M., 1997).

The concept of ecological footprinting was first developed by Rees and Wackernagel in the early 1990s as an elaboration of the "carrying capacity" concept (Wackernagel and Rees, 1996). *The ecological footprints of fuels* consists of three main components: the area needed for energy production, the area needed to sequester emissions of greenhouse gases, and the area needed for the safe deposit of nitrogen and sulphur (Holden and Hoyer, 2005).

Using conventional technology, the potential for reducing the ecological footprint by 2010 is substantial: a 22 per cent reduction can be achieved for petrol cars and more than 35 per cent for diesel-fuelled cars.

Compared to the reference chain (conventional raw oil based petrol used in an internal combustion engine) presented energy chains display differences in the size of the ecological footprints, however it seems that differences between the groups are more important (see Table 1).

Table 1: Comparison of energy chains

Energy chain	**Ecological footprint compared to the reference energy chain (%)**[a]
The hydropower-based energy chains	75% reduction
The natural gas based energy chains	45-75% reduction
The raw oil based energy chains	15-30% reduction
The biomass (wood)-based energy chains	0-50% increase

[a] The reference energy chain (raw oil–petrol–conventional power train in 2010).

Hydropower has a very low ecological footprint, but is not a global resource with sufficient volumes to support the ever-increasing transport systems. Natural gas also has a low ecological footprint, but does not have these resource volume limitations; at least it will not for several decades. But it is not a renewable energy resource, and it does not fulfil the long-term requirements of a sustainable energy system. Biomass is globally available in large volumes and is a renewable resource, but would lead to unacceptable increases in ecological footprints if used extensively.

Therefore, it seems that through the development of methodology to estimate the ecological footprint of fuels, the combined use of resources, the substitution of less environmentally harmful fuels, and reduced transportation, long-term sustainability objectives can be met.

2.2 Air pollution from traffic

Energy use in the transportation sector significantly contributes to air pollution, acidification and climate change through emissions of carbon monoxide (CO), nitrogen oxides (NO_x), sulphur oxide (SO_2), particulate matter (PM), hydrocarbons (HC) and volatile organic compounds (VOC). These pollutants are released during the combustion of fossil fuels. Additional emissions of particulate matter and PM-bound compounds could originate from friction acting on individual components such as tyres and brakes.

The secondary air pollutants are substances created in chemical reactions among primary pollutants. For example, NOx and HC in the presence of sunlight create ozone (O_3). Ozone is the main component of photochemical smog and the major air pollutant in urban areas in temperate regions.

Carbon monoxide (CO) is one of the most common and widely distributed air pollutants. In the human body, it reacts readily with haemoglobin to form carboxy-haemoglobin (COHb). Global background concentrations of carbon monoxide range between 0.06 mg/ m^3 and 0.14 mg/ m^3 (0.05-0.12 ppm).

Anthropogenic emissions of carbon monoxide originate mainly from incomplete combustion of carbonaceous materials. The largest proportion of these emissions is produced by exhausts of internal combustion engines, especially by motor vehicles with petrol engines. Other common sources include various industrial processes, power plants using coal, and waste incinerators. Petroleum-derived emissions have greatly increased during the past few decades (Cullis, C.F. and Hirschler, M.M., 1989). Some widespread natural non-biological and biological sources, such as plants, oceans and the oxidation of hydrocarbons, give rise to the background concentrations outside urban areas.

The ambient concentrations measured in urban areas depend greatly on the density of combustion-powered vehicles, and are influenced by topography and weather conditions. In the streets, the carbon monoxide concentration varies greatly according to the distance from the traffic; it is also influenced by topography and weather conditions. In general, the concentration is highest on the leeward side of the "street canyon", and there is a sharp decline in the concentration from pavement to rooftop level (Rudolf, W., 1994).

In the urban traffic environments of large European cities, 8-hour average carbon monoxide concentrations are generally lower than 20 mg/ m^3 (17 ppm), with short-lasting peaks up to 60 mg/ m^3 (53 ppm). Carbon monoxide levels have a close quantitative and temporal association with the levels of other primary exhaust pollutants, such as nitrogen monoxide and volatile organic compounds (Derwent, R.G. et al., 1995; Dor, F. et al., 1995).

Carbon monoxide concentrations inside vehicles are generally higher than those measured in ambient outdoor air. Traffic patterns, car model and maintenance, vehicle ventilation conditions and season are all factors that affect carbon monoxide levels inside cars (Chan, C.-C. et al., 1991; Flachsbart, P.G., 1995; Zhang, Y. et al., 1995). It has been shown that bicycle-commuters in Southampton in the UK are exposed to mean concentrations of 6.1-20.5 mg/m^3 (5.3-17.9 ppm), with short peak values as high as 71 mg/ m^3 (62 ppm) (Bevan, M.A.J. et al., 1991). In the Netherlands, in Amsterdam's inner city, the corresponding mean concentrations encountered by cyclists were much lower: 0.6-4.1 mg/m^3 (0.5-3.6 ppm) (Van Wijnen, J.H. et al., 1995).

Carbon monoxide concentrations measured at fixed-site monitoring stations seem to inaccurately reflect short-term personal exposures of various urban population groups. Cortese & Spengler (Cortese, A.D. and Spengler, J.D., 1976) reported that, among a group of commuters living in the Boston metropolitan area in the US, the 1-hour average personal concentrations measured by portable technology were 1.3-2.1 times the concentrations measured by fixed-site equipment. Furthermore, in Washington DC and Denver in the US (Akland, G.G. et al., 1985), and also in Helsinki in Finland (Alm, S. et al., 1994), there was no significant association between personal 1-hour exposure levels and the corresponding carbon monoxide concentrations measured at fixed-site monitoring stations. However, the fixed-site monitoring data may reflect population exposures somewhat better over longer sampling periods, for example 8 hours (Cortese, A.D. and Spengler, J.D., 1976; Akland, G.G. et al., 1985; Liu, J.-J. et al., 1994).

In underground and multi-storey car parks, road tunnels, enclosed arenas and other indoor microenvironments, in which combustion engines are used under conditions of insufficient ventilation, the mean levels of carbon monoxide can rise above 115 mg/m^3 (100 ppm) for several hours, with short-lasting peak values that can be much higher.

Conversion factors:

1 ppm = 1.145 mg/m^3
1 mg/m^3 = 0.873 ppm

Nitrogen oxides (NOx) emissions are mainly produced from atmospheric nitrogen and oxygen as a result of the combustion of fuels under high pressures and at high temperature. Car traffic is a significant source of nitrogen oxides air pollution. With the installation of catalytic converters in cars, a reduction in nitrous oxides emissions is achieved; it is only limited, however, due to the simultaneous growth in the number of cars being driven. Emissions from industrial processes (not associated with fuel combustion, e.g. the production of nitric acid) have only minor importance in Europe (EEA, 1996). *Fuel* NO_x and *thermal* NO_x are two of the main types of chemical reactions among the most important formation mechanisms for nitrogen oxides during the combustion of fossil fuels.

Fuel NO_x. During combustion, the nitrogen chemically associated with the fuel (apart from the molecular nitrogen contained for example in natural gas) converts to amines and cyanides, which then combine together with oxygen to form nitrogen oxides. This "fuel NO_x" formation is a function of the fuel's nitrogen content, as well as of the burner type and firing mode that is used. In coal, fuel nitrogen content typically varies between 0.5 and 2.0 per cent (by weight). The figure is less than 1 per cent for oil, and in natural gas the nitrogen content is negligible. Because of the simultaneous reverse reaction (i.e. the formation of nitrogen from nitrogen oxides), only between 5 and 25 per cent of the total fuel nitrogen is typically converted to NO_x. Only a small amount of NO_x is produced from nitrogen in fuel

Thermal NO_x. Generation of thermal NO_x occurs due to the mechanism discovered by Zeldovich, in which nitrogen and oxygen from the air combine to form NOx under high temperature:

$$N_2 + O_2 \rightarrow 2NO;$$
$$N + O_2 \rightarrow NO + O\,.$$

This formation process usually becomes important at temperatures above 1400 degrees C (Rentz et al., 1987), temperatures which are generally exceeded in most combustion processes. At higher temperatures, thermal NO_x generation increases exponentially. It depends also on how long the combustion air remains in the combustion chamber, and on the availability of excess air.

The largest fraction of nitrogen oxides emissions are emitted as NO (monoxide) and are oxidised in the atmosphere to NO_2 dioxide. More and more interest has been shown recently in the generation of N_2O because of its contribution to the global greenhouse effect. However, all species of nitrogen oxides are usually converted to NO_2.

Sulphur dioxide (SO_2) is derived from the combustion of fossil fuels containing sulphur and is a major air pollutant in many parts of the world. Oxidation of sulphur dioxide, especially on the surface of particles in the presence of metallic catalysts, leads to the formation of sulphurous and sulphuric acids. Neutralisation, by ammonia, leads to the production of bisulphates and sulphates.

As a result of changes in sources, annual mean levels of sulphur dioxide in the major cities of Europe are largely below 100 $\mu g/m^3$. Daily mean concentrations have also fallen, and are now generally below 500 $\mu g/m^3$. Natural background concentrations of sulphur dioxide in rural areas of Europe are generally below 5 $\mu g/m^3$.

Sulphur dioxide emissions are not important with regard to car traffic, since petrol does not contain much sulphur; however, they are important with respect to sea transport, which uses oil. The amount of sulphur in any fuel contains depends on where and how it was formed.

Historically, sulphur dioxide and particulate matter derived from the combustion of fossil fuels have been the main components of air pollution in many parts of the world. The most serious problems have been encountered in large urban areas where coal is used for domestic heating purposes, or for poorly controlled combustion in industrial installations. In such situations the complex of pollutants has generally been considered collectively, drawing on findings from epidemiological studies done several decades ago in areas formerly heavily polluted. Guidelines developed in this way had been related to averaging times of 24 hours for acute effects and one year for chronic effects (Air quality guidelines for Europe, 1987).

Separate attention has been given to sulphur dioxide alone, based largely on findings from controlled human-exposure studies. These allow guidelines to be developed in terms of shorter averaging periods in the order of one hour. These are relevant to exposures to peak concentrations that may arise from sources such as burning coal or heavy oil, whether or not they are accompanied by substantial concentrations of particulates.

Conversion factors:

1 ppm (20ºC, 1013hPa) = 2860 $\mu g/m^3$
1 mg/m^3 = 0.350 ppm

Particulate matter (PM) in urban air often strongly depends on the influence of regionally transported particulates and gaseous pollutants. This phenomenon causes low spatial variability of concentrations. However, there is evidence that mass concentrations of fine particulate matter tend to be higher, as compared to background values, in areas close to sources of strong pollution. Roads and streets with intense motor vehicle traffic represent one of these sources, producing both primary and secondary particles, including those emitted with exhaust gasses and those resuspended from the pavement. The emission of particulate matter is associated primarily with diesel engines, which produce 30 to 70 times more emissions than petrol engines.

Atmospheric particles originate from a variety of sources and possess a range of morphological, chemical, physical, and thermodynamic properties. The airborne particles are of many kinds: resuspended soil particles, smoke from power generation, photochemically formed particles, atmospheric clouds of water droplets, etc. These airborne particles are all examples of aerosols. The aerosols include a wide range of phenomena, such as dust, fume, smoke, mist, fog, haze, clouds, and smog. Atmospheric aerosols are also referred to as suspended particulate matter. Particles less than 10 μm in diameter have very low sedimentation speed under gravity, and they may remain in the air for several days before eventually being washed out by rain or impacted out onto vegetation or buildings. They are a significant environmental pollutant, being responsible for loss of visual range, soiling of surfaces and human health effects. Unlike other criteria pollutants (O_3, CO, SO_2, NO_2, etc.), particulate matter is not a specific chemical entity but is a mixture of particles from different sources and of different sizes, compositions, and properties. An accurate understanding of their concentrations, size distributions, and chemical composition is therefore necessary, but for various reasons this is quite hard to achieve.

Fig. 3: Schematic of an urban particle size distribution with major sources and transformation processes (Source: Whitby and Cantrell, 1976)

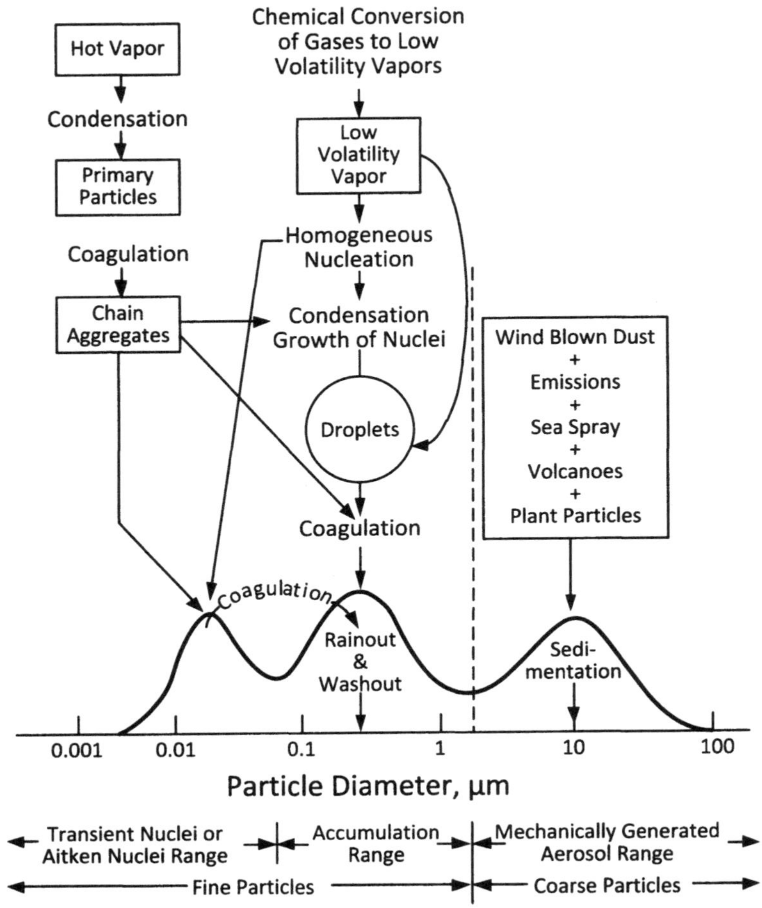

Research on particle inhalation processes has led to the identification of the aerodynamic size classes of the particulate matter. The International Standards Organization (ISO 7708) has defined four main categories of airborne particles, according to their ability to penetrate the human respiratory system: inhalable (total), thoracic (median diameter 10 μm), respirable (4 μm), and high-risk respirable (2.5 μm). The thoracic and respirable categories were applied to the atmospheric particulate matter and currently serve as air quality standards in many countries worldwide. The US EPA defines the PM10 category as particles measured by a sampler that contains a size fractionator (classifier) designed with an effective cut-off point (50% collection efficiency) of 10 μm aerodynamic diameter. Similarly, PM2.5 is defined as particles measured by a sampler that contains a size fractionator designed with an effective cut-off point (50% collection efficiency) of 2.5 µm aerodynamic diameter. The PM2.5 standard was based primarily on epidemiologic studies using concentrations measured with PM2.5 samplers as an exposure index. The PM2.5 is usually referred to as a "fine" fraction of the aerosol, while particles which fall in the categories higher than 2.5 µm are referred to as the "coarse" fraction (Figure 3). The fine fraction consists of three aerosol modes: nucleation, Aitken and accumulation (Figure 4). The nucleation and Aitken mode are invisible in mass distribution, because very small diameter particles do not contain significant mass. The fraction of PM2.5 that is smaller than 0.1 µm is also referred to as "ultrafine" particles.

Fig. 4: An idealised size distribution that might be observed in traffic, showing fine and coarse particles and the nucleation, Aitken, and accumulation modes that comprise fine particles. Also shown are the major formation and growth mechanisms of the four modes of atmospheric particles (US EPA, 2004).

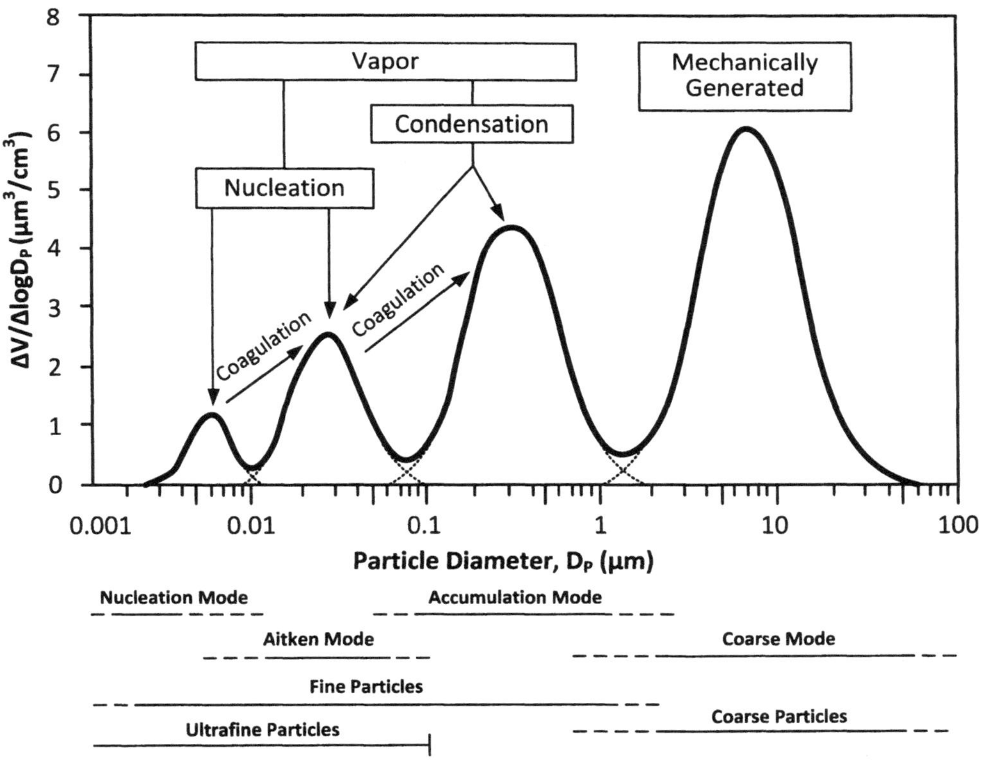

The fine and coarse fractions are distinctly different not only in terms of particle size, but also by formation sources and chemical composition. The coarse fraction is generated by the break-up of droplets or solids, and consists of suspended crustal material, road wear debris, fly ash, biological particles (pollen and fungal spores). The fine fraction originates from combustion, high-temperature processes, and atmospheric reactions. It is formed by nucleation, condensation and coagulation processes, and reactions of gases in or on particles. It is composed of soluble ions (sulphate, nitrate, ammonium, hydrogen), a large variety of organic compounds, elemental carbon, and trace metal compounds (Martuzevicius, D., 2005).

The chemical mechanisms responsible for the formation of carbonaceous particles in diesel engines are not well established, but are thought to involve the intermediate formation of polycyclic aromatic hydrocarbons. Roughly 10-40 per cent of particulate emissions from diesels are extractable into organic solvents (National Research Council, 1982). In a typical profile, 50 per cent of the extract is composed of aliphatic hydrocarbons with 14-35 C atoms and alkyl-substituted benzenes; 4 per cent are PAHs and heterocycles; and about 6 per cent are PAH oxidation products including a small fraction of nitro-PAHs.

Particulate matter is also formed in internal combustion engines as a result of the incomplete combustion of gasoline with contributions from engine oil. The particles consist essentially of a solid carbon core with a coating of organic compounds, sulphate, and trace elements. The composition of PAHs, oxy-PAHs and their alkyl homologues in tailpipe emissions from petrol-fuelled vehicles is similar to that produced in diesel engines (Behymer and Hites, 1984).

The concentrations of particulate matter were found by Hammerle et al. (1992) to roughly correlate with the total hydrocarbons in the exhaust of petrol-fuelled vehicles. Emission factors for particulate matter in the exhaust of petrol-fuelled vehicles range from 0.011 g/km for light-duty vehicles to 0.12 g/km for heavy-duty petrol vehicles, and from 0.23 g/km in the exhaust of diesel passenger vehicles to 1.20 g/km for heavy-duty diesel vehicles (Radwan, 1995). These values are based on characteristics of the motor vehicle fleet in 1990.

Volatile organic compounds (VOCs) are carbon-based compounds (with 2-10 carbon atoms) that have vapour pressures high enough to significantly vaporize and enter the atmosphere. Many different kinds of VOCs can be found in the air: alkanes, alkenes, alkynes, halogenated hydrocarbons, aromatic hydrocarbons, terpenes, aldehydes, ketones and alcohols. Some of these compounds are toxic or carcinogenic, and therefore there are limit values for their concentrations in the air (US EPA, 2005; EU, 2000).

VOCs affect atmospheric chemistry in many ways. In the atmosphere they are oxidized by hydroxyl radicals, ozone, nitrate radicals and halogens (Cl, Br, I) and, in addition, some of them can be photolysed. In the presence of nitrogen oxides they contribute to ozone formation in the lower troposphere (Atkinson, 2000):

$$VOC + NO_x + \text{sunlight} \rightarrow O_3 + \text{“other products”}$$

In the reactions of VOCs, water soluble hydroperoxides, carbonyls and acids are produced; VOCs therefore make a contribution to the organic content and acidity of precipitation (Kawamura et al., 2001). One important aspect is that the reaction products of VOCs may also take part in the formation and growth of new particles, with possible climate and health consequences (Griffin et al., 1999; Hoffmann et al., 1997).

Globally, biogenic emissions of VOCs exceed those of anthropogenic origin (Müller, 1992). However, in urban areas the contribution of biogenic VOCs is much lower. Anthropogenic VOC sources include combustion processes, the use of fossil fuels, solvents, industrial production processes and biological processes (Friedrich and Obermeier, 1999). Whereas VOC emissions from combustion sources (e.g. traffic and wood combustion) mainly contain pure hydrocarbons, organic solvents and their vapours also consist of oxygenated HCs such as alcohols, carbonyls and esters.

Traffic and traffic-related sources are known to be a major source of non-methane hydrocarbons (NMHCs, i.e. alkanes, alkenes, alkynes and aromatic HCs) in urban areas (Friedrich and Obermeier, 1999; Watson et al., 2001), but in residential or industrial areas other sources may also be important. In Nordic countries the use of wood as a fuel has increased lately (Haaparanta et al., 2003; Hedberg et al., 2002) and wood combustion is known to emit several different VOCs (i.e. NMHCs, halogenated hydrocarbons and oxygenated hydrocarbons) and other air pollutants (McDonald et al., 2000). For the lightest alkanes, natural gas emissions may also be important (Fujita, 2001). Although ethene is a major constituent of the VOC emissions from traffic and from wood combustion (Schauer et al., 2002 and McDonald et al., 2000), it is also a plant hormone and is emitted by plants, soils and oceans (Fall, 1999). In addition to this, terpenes (isoprene and monoterpenes) have mainly biogenic sources.

In some large cities, concentrations of VOCs can be very high compared with those in remote areas. In a recent study in Mumbai, India (Srivastava et al., 2006), the annual average of benzene concentrations during rush hours in commercial areas and at traffic intersections were 127 g/m^3 and 348 g/m^3 respectively, and even in residential areas the average concentration was over 40 g/m^3. In some European cities quite high concentrations have also been measured: for example, in a medium-sized Greek city, Ioannina, the annual average benzene concentrations measured at different locations in 2003/2004 were between 10 and 40 g/m^3 (Pilidis et al., 2005). In measurements by Hopkins et al. (2002), taken on a ship in the Arctic during August 1999, much lower benzene concentrations were found, averaging about 0.15 g/m^3; in the studies by Kato et al. (2001) in December 1999 meanwhile, on a remote island in the Pacific Ocean, the average concentration of benzene was measured at about 0.38 g/m^3. High differences in concentrations are also measured for most other VOCs. The life-time of benzene is quite long and therefore it can be transported far from its emission sources. This is not the case for the more reactive compounds: taking for example a rural forested site in Central Finland, even in winter, when the highest concentrations of NMHCs are measured, the concentrations of most other aromatic hydrocarbons are below detection limits (Hakola et al., 2003). In rural and remote areas NMHCs show a very clear seasonal cycle; in the Northern Hemisphere, the highest concentrations are measured in winter and the lowest in summer (Hakola et al., 2006; Gautrois et al., 2003). Winter maxima and summer minima of NMHCs are also observed in urban areas (e.g. Morikawa et al., 1998, Sahu and Lal, 2006). For biogenic hydrocarbons and some carbonyls, the cycle works the other way round: maximum concentrations are observed in summer, while minima occur in winter (Hakola et al., 2003 and Solberg et al., 1996). In Western Europe, emissions of VOCs have been decreasing since the early 1990s, and a decreasing trend in ambient concentrations has also been observed in Central Europe (Solberg et al., 2002). However, in remote areas of Finland for example, no clear decrease in concentrations has been found for most of the compounds; for some long-living compounds (ethane and propane) increasing trends have even been observed (Hakola et al., 2006).

2.3 Climate change

The United Nations' Framework Convention on Climate Change is the first binding international legal instrument aimed at establishing greenhouse gas concentrations in the atmosphere at a level that would prevent dangerous anthropogenic interference with climatic systems. According to the Kyoto Protocol, the current target for European states is to reduce CO_2-equivalent emissions to 12 per cent of the 1990 level by the year 2012. CO_2-equivalent includes carbon dioxide CO_2, nitrous oxide N_2O, methane CH_4, sulphur hexafluoride SF_6, hydrofluorocarbons HFCs, and perfluorocarbons PFCs. These are the gases we should manage.

Many sectors, including transportation, are responsible for the emission of greenhouse gases. Usually, the contribution of the transport sector is estimated from the statistical data on fuel consumption or the mobility of citizens living within the local authority area. Increased kilometres travelled reflect increased local contribution to global climate change. The different aspects (and the related units of measurements) that contribute to defining the general model of mobility of each citizen include:

- The number of trips that, on average, each citizen makes during the day, where "trip" indicates a journey with a starting-point and a destination (number of daily trips per capita);
- The reason for these trips and their regularity during the week, allowing for the trips to be classifies as either "systematic" or "unsystematic" (% of systematic trips compared with unsystematic ones);
- The average distance covered by each citizen during the day (km/per capita);
- The time taken by each citizen for his/her trips;
- Modes of transport used for the trips and/or for the different distances associated with each trip (% relating to the different modes of transport considered).

Kilometres travelled per capita are estimated using data from personal, public, and commercial modes of transportation. If such data is not available it may be preferable to multiply the size of the vehicle fleet by the estimated average fuel consumption. Fuel consumption is always available on the national level, but it is more difficult to calculate for a city, since a lot of fuel purchased in the city may be used for inter-city travel. Fuel purchase within a city forms a part of the city's product; therefore evaluation of average fuel consumption could be based on revenues from motor fuel taxes (Kliucininkas L., 2003).

While the estimation of fuel consumption includes fossil fuel used by motor vehicles, it does not reflect mobility by electric-driven transport (rail or trolley bus), nor does it include electricity generated for transport. It is conceptually difficult to separate motor vehicle fuel purchased for urban areas from that purchased for non-urban areas. These factors should be considered when interpreting the estimation results.

2.4 Noise

Noise is probably the most obvious impact of urban transport. Excessive noise levels (65 dB(A) and higher) are damaging to public health. It contributes to high blood pressure and cardiovascular diseases. Some 65 per cent of the population of Europe is exposed to environmental noise levels above 55 dB(A), the level at which noise may cause annoyance and sleep disturbances. About 17 per cent are exposed to levels above 65 dB(A), at which point the level impact of noise is noticeable. Approximately 1.4 per cent of the population of Europe is exposed to levels above 75 dB(A). Exposure to traffic noise differs between countries (Tengstrom, E. and Thynell, M., 1997).

Silence is an important resource currently widely intruded upon by the transportation sector. Noise can induce physiological reactions. People react differently to traffic noise and it is therefore difficult to define absolute limits above which noise can be regarded as a disturbance. Noise levels can differ significantly in the same general area and there can be a large variation over days, weeks and years. Continuous exposure to noise can lead to weakening of the auditory system and sleeping disorders. Noise also has negative effects on wildlife.

Each mode of transport emits noise in different ways and the most important source of noise may change depending on the situation, e.g. speed and acceleration. Additionally, the frequency spectrum emitted by each mode differs, which is important as human noise perception varies for the different frequencies.

For road modes, there are three main sources of noise: (i) the engine, with its combustion processes; (ii) the mechanics, at the interface between the tyre and the road which generates the rolling noise; and (iii) the wind noise, depending on the aerodynamic characteristics of the vehicle. Which of these is the dominant source of noise depends very much on the driving speed. At low speeds up to about 30 km/h, the engine noise is most significant. As speed increases, the tyre-road rolling noise becomes the most significant source, and finally at speeds of 60 to 80 km/h, the wind noise becomes the most significant source. These distinctions are true for driving at more or less continuous speeds.

If the vehicle is accelerating, engine noise can also be a significant source at much higher speeds than 30km/h. For large vehicles too, engine noise is a significant source, even when travelling at higher speeds. In addition to being louder, lorries and buses also emit more noise in low frequency bands than passenger cars. It is more difficult to insulate buildings against low frequency bands, which means the share of lorries and buses has to be carefully considered in noise assessments. The impact of acceleration on noise emissions also needs to be considered. Traffic situations requiring acceleration, such as traffic lights, pedestrian crossings and steep uphill roads, usually generate more transport noise.

There are two main approaches to measuring transport noise immissions (a technical term for the noise level imposed on a given location by a given source). In the first, noise is measured directly at the reception point by measuring the noise immissions, e.g. with a microphone or similar technical equipment. In the second, noise is calculated by computer models that use transport characteristics and local parameters as inputs, in order to derive noise emissions from the source and the resulting noise immissions at the reception point. To get comparable and meaningful results, direct measurement of noise requires standardised equipment and procedures.

The basic approaches to noise reduction focus on technological improvements, land use and traffic planning. Other approaches involve limiting the transmission of noise and reducing noise at the perception location (e.g. through the installation of noise protection barriers and windows, or insulation). Another important approach looks for significant changes in societal values and lifestyle.

The effect of technological measures at the noise source always takes time, since the new technology has to be installed in the vehicle fleet, either through upgrading the existing vehicle fleet or by waiting for its natural renewal. Comprehensive city and land-use planning could play a key role in reducing urban noise levels. The other two approaches to reducing noise are limiting transmission and reducing perception, both of which require construction (e.g. noise protection walls, tunnels, windows), which cost considerable amounts of money. Voluntary behavioural change also requires longer periods of time, because it has to be fostered by education and long-term changes in lifestyle. It seems that only if the whole package of measures were implemented, sufficient achievement of noise targets would be guaranteed.

2.5 Space intrusion

Increased demand for transport is already creating enormous pressure for new infrastructure. Many of these transport facilities require very large amounts of land, both for their own internal operations and for the external transport links that have to be provided. A fundamental question is, can the environment and society afford to provide sites of the scale required by the transport industry?

Urban morphology has become increasingly complex over time, particularly throughout the latter half of the twentieth century. High rates of urbanisation, accompanied by rapid suburbanisation, have transformed the monocentric city – the dominant urban form for several millennia – into one that is polycentric. The continuous decentralisation of population and employment from the core to the periphery of cities is well documented in the literature under the labels of "counter-urbanisation" and "exurbanisation" (Marmaras, E. and Kanaroglou, P., 1996).

Over recent decades, many large European cities have grown in a scattered way. The current trend towards a decentralisation of population and employment has led to longer trips and an increasing dependence on cars, exerting great pressure on the environment. The environmental, economic, and social problems associated with urban mobility patterns represent the core of the debate on the possible influence of urban form on mobility patterns. In most European countries, transportation planning is carried out independently from land-use planning. Policy formulation and planning in the transportation sector, usually the domain of engineers and economists, falls under the jurisdiction of national governments. The prevailing idea among transportation planners, then, is that land is developed in the open market, and that their role is to develop infrastructure that will accommodate the evolution of land development in the best possible way.

The 1990s brought a change to this attitude, driven primarily by environmental concerns. The transportation sector was identified as a major contributor of harmful environmental emissions in urban areas, and many cities have now set targets for emission reductions. The empirical literature suggests that improvements in automobile engine technologies will not be sufficient in meeting desired urban sustainability goals (Anderson et al., 1996a, 1996b). Reductions in the frequency and length of trips will be necessary. This realisation led to the conclusion that land-use policies that will bring the origins and destinations of motorised trips closer to each other need to be designed and implemented. The circular nature of the impacts between transportation and land use has been captured effectively by Wegener (1995), who argues for the integration of land-use and transportation planning.

Articulating the concept of a city and the processes that take place in it is not necessarily straightforward. Three useful terms in this context are: "urban form", "urban interaction" and "urban spatial structure". "Urban form" may be defined as the spatial configuration of fixed elements within a metropolitan area. "Urban interaction" refers to the flows of people, goods and information among different locations in the city. "Urban spatial structure" is a comprehensive concept. According to Bourne (1982), it consists of three elements: urban form, urban interaction and a set of organising principles that define the relationship between the two. The key point here is that urban form exerts a profound influence on the transportation flows within the city, but does not determine them completely. One can envisage, for example, two very different commuting patterns that might overlay the same urban form. The first is one in which people commute to places of work close to their residences; in the second, people commute longer distances, by passing nearby employment districts for ones farther away. In the first case, workers locate themselves close to their place of work, making efficient commuting the organising principle. In the second, a different factor – such as proximity to relatives or people of the same ethnic origin and status – determines residential location. In reality, within a complex urban system several organising principles work concurrently.

The urban system is also a dynamic entity, its spatial structure evolving over time. A host of factors are responsible for this evolution, not least of which is the behaviour of the different institutions – households, firms and public institutions – that operate in the system. The locational decisions of these institutions can affect urban form. For example, empirical evidence suggests that the locational decision of a household is associated with its car ownership decision, and ultimately with the number and length of trips a household takes per week. Thus, the behaviour of these institutions is intimately related to the organising principles that give rise to urban spatial structure.

Travel patterns in the metropolitan areas of most developed countries are increasingly car-dependent, and less subject to patterns characterised by trips from the periphery to the main centre or just outside. Car dependency has become a political priority for Western European governments (OECD, 1995) with regard, among other factors, to the environmental problems caused by this mobility pattern. The academic debate has mainly focused on the impact of urban spatial structure on travel patterns. However, it is not clear that causality works only in that direction. The debate on the relationship between intra-urban mobility (travel patterns) and urban form (land use characteristics) was reactivated by the publication of Newman's and Kenworthy's (1989) research. Banister (1998) suggests that there are two important reasons why population density may reduce the ecological impact of mobility. First, higher density patterns reduce average distances between home and place of work; second, high densities may be more amenable to public transport supply.

Much research on the influence of urban form on mobility has been done using indicators of travel patterns such as travel distance and modal choice. Some authors have nonetheless used a more interesting indicator: energy consumption, a composite measure of travel distance, modal choice, and journey frequency. Muniz and Galindo (2005) offer the *ecological footprint* calculation for commuting in the Barcelona Metropolitan Region. The objectives of the research are threefold: first, measuring the evolution of commuting trips (length and modal split); second, calculating the ecological footprint of commuting and its variation between 1986 and 1996; and third, determining the effect of urban form and other factors on commuting eco-footprint. One of the most controversial ideas in the debate on urban sustainability is that urban sprawl causes problems of ecological stress. This widespread assumption has been tested by measuring the ecological footprint left by commuters in the 163 municipalities of the Barcelona Metropolitan Region. The following regressors were used as determinants of the ecological footprint: population density, accessibility, average household income, and job ratio. The results confirm that urban form appears as the main determinant of ecological footprint variation among the municipalities of Barcelona Metropolitan Region. Despite certain inherent limitations with this indicator (e.g. technology and level of vehicle occupancy are not necessarily constant over time), the study provides interesting results.

References

Air quality guidelines for Europe (1987). Copenhagen, WHO Regional Office for Europe (WHO Regional Publications, European Series, no. 23).

Akland, G.G. et al. (1985). Measuring human exposure to carbon monoxide in Washington, D.C. and Denver, Colorado, during the winter of 1982-1983. *Environmental science and technology*, 19, pp. 911-918.

Alm, S. et al. (1994). Personal exposures of preschool children to carbon monoxide: roles of ambient air quality and gas stoves. *Atmospheric environment*, 28, pp. 3577-3580.

Anderson, D. (1997). Progress towards energy sustainability in OECD countries, Rio +5 report.

Anderson, W.P., Kanaroglou, P.S. and Miller, E.J. (1996a). Urban form, energy and the environment: A review of issues, evidence and policy, *Urban Studies*, 33, pp. 7-35.

Anderson, W.P., Kanaroglou, P.S., Miller, E.J. and Buliung, R.N. (1996b). Simulating automobile emissions in an integrated urban model, *Transportation Research Record*, 1520, pp. 71-80.

Atkinson, R. (2000). Atmospheric chemistry of VOCs and NOx. *Atmospheric Environment*, 34, 2063-2101.

Banister, D. (1998). *Sustainable Development and Transport. Report for the Bundesforschungsanstalt für Landeskunde und Raumordnung*. The Urban 21 Project.

Behymer, T.D., Hites, R.A. (1984). Similarity of some organic compounds in spark-ignition and diesel engine particulate extracts. *Environ. Sci. Technol.* 18, pp. 203-206.

Bevan, M.A.J. et al. (1991). Exposure to carbon monoxide, respirable suspended particulates, and volatile organic compounds while commuting by bicycle. *Environmental science and technology*, 25:788-791.

Bourne, L.S. (1982). Urban spatial structure: An introductory essay on concepts and criteria, in L.S. Bourne (ed.), *Internal structure of the city*, New York: Oxford University Press, 2nd edition, 28-45.

Cameron, I., Kenworthy, J.R. and Lyons, T.J. (2003). Understanding and predicting private motorized urban mobility. *Transportation Research Part D* 8, pp. 267-283.

Chan, C.-C. et al. (1991). Driver exposure to volatile organic compounds, CO, ozone, and NO_2 under different driving conditions. *Environmental science and technology*, 25, pp. 964-972.

Cortese, A.D. and Spengler, J.D. (1976). Ability of fixed monitoring stations to represent personal carbon monoxide exposure. *Journal of the Air Pollution Control Association*, 26:1144-1150.

Cullis, C.F. and Hirshler, M.M. (1989). Man's emissions of carbon monoxide and hydrocarbons into the atmosphere. *Atmospheric environment*, 23:1195-1203.

Derwent, R.G. et al. (1992). Analysis and interpretation of air quality data from an urban roadside location in central London over the period from July 1991 to July 1992. *Atmospheric environment*, 29, pp. 923-946.

Dor, F. et al. (1995). Exposure of city residents to carbon monoxide and monocyclic aromatic hydrocarbons during commuting trips in the Paris metropolitan area. *Journal of the Air & Waste Management Association*, 45, pp. 103-110.

EU (2000). Directive (2000/69/EC) of the European Parliament and of the Council of 16 November 2000 relating to limit values of benzene and carbon monoxide in ambient air, *Official Journal of the European Communities* no. L 313/12, 13.12.2000.

Fall, R. (1999). Biogenic emissions of volatile organic compounds from higher plants. In Hewitt C.N. (ed.): *Reactive hydrocarbons in the atmosphere*, Academic Press, London, pp. 1-39.

Friedrich, R. and Obermeier, A. (1999). Anthropogenic emissions of volatile organic compounds. In Hewitt C.N. (ed.): *Reactive hydrocarbons in the atmosphere*, Academic Press, London, pp. 1-39.

Fujita, E. (2001). Hydrocarbon source apportionment for the 1996 Paso del Norte Ozone Study. *The Science of the Total Environment*, 276, pp. 171-184.

Gautrois, M., Brauers, T., Koppmann, R., Rohrer, F., Stein, O. and Rudolph, J. (2003). Seasonal variability and trends of volatile organic compounds in the lower polar troposphere. *Journal of Geophysical Research*, 108, pp. 4393-4408.

Griffin, R.J., Cocker III, D.R., Seinfeld, J.H. and Dabdub, D. (1999). Estimate of global atmospheric organic aerosol from oxidation of biogenic hydrocarbons. *Geophysical Research Letters*, 26, pp. 2721-2724.

Haaparanta, S., Myllynen, M. and Koskentalo, T. (2003). Small-scale wood combustion in the Helsinki Metropolitan Area. *Pääkaupunkiseuden julkaisusarja PJSB* 18, pp. 1-44. Helsinki Metropolitan Area Council.

Hakola, H., Hellén, H. and Laurila, T. (2006). Ten years of light hydrocarbon (C2-C6) concentration measurements in background air in Finland. Accepted to *Atmospheric Environment*.

Hakola, H., Tarvainen, V., Laurila, T., Hiltunen, V., Hellén, H. and Keronen, P. (2003). Seasonal variation of VOC concentrations above a boreal coniferous forest. *Atmospheric Environment*, 37, pp. 1623-1634.

Hammerle, R.H., Korniski, T.J., Weir, J.E., Chladek, E., Glerczak, C.A., Chase, R.E., Hurley, R.G. (1992). Effect of mileage accumulation on particulate emissions from vehicles using gazoline with methylcyclopentadienyl manganese tricarbonyl. Society of Automotive Engineers; *SAE technical paper* no. 920731.

Hoffman, T., Odum, J.R., Bowman, F., Collins, D., Klockow, D., Flagan, R.C. and Seinfeld, J.H. (1997). Formation of organic aerosols from the oxidation of biogenic hydrocarbons. *Journal of Atmospheric Chemistry*, 26, pp. 189-222.

Holden, E. and Hoyer, K.H. (2005). The ecological footprints of fuels. Transportation Research Part D: *Transport and Environment*, vol. 10, iss. 5, pp. 395-403.

Hopkins, J.R., Jones, I.D., Lewis, A.C., McQuaid, J.B. and Seakins, P.W. (2002). Nonmethane hydrocarbons in the Arctic boundary layer. *Atmospheric Environment*, 36, pp. 3217-3229.

Kato, S., Pochanart, P. and Kajii, Y. (2001). Measurements of ozone and nonmethane hydrocarbons at Chichi-jima island, a remote island in the western Pacific: longrange transport of polluted air from the Pacific rim region. *Atmospheric Environment*, 35, pp. 6021-6029.

Kawamura, K., Steinberg, S., Ng, L. and Kaplan, I.R. (2001). Wet deposition of low molecular weight mono- and di-carboxylic acids, aldehydes and inorganic species in Los Angeles. *Atmospheric Environment*, 35, pp. 3917-3926.

Kliucininkas, L. (2003). *Local sustainability indicators. Urban Environmental Management* vol. 4. The Baltic University Press, pp. 13-26.

Liu, J.-J. et al. (1994). Predicting personal exposure levels to carbon monoxide (CO) in Taipei, based on actual CO measurements in microenvironments and a Monte Carlo simulation method. *Atmospheric environment*, 28, pp. 2361-2368.

Marmaras, E. and Kanaroglou, P. *Integrated urban transportation and land-use models: an concise overview of the state of art*. Department of Geography, University of the Aegean. P. 6 (typescript).

Martuzevicius, D. (2005). *Variations of the traffic-related fine aerosol fraction in an urban airshed*. Doctoral dissertation, Kaunas University of Technology.

McDonald, J.D., Zielinska, B., Fujita, E.M., Sagebiel, J.C., Chow, J.C. and Watson, J.G. (2000). Fine particle and gaseous emission rates from residential wood combustion. *Environmental Science and Technology*, 34, pp. 2080-2091.

Morikawa, T., Wakamatsu, S., Tanaka, M., Uno, I., Kamiura, T., and Maeda, T. (1998). C2-C5 hydrocarbon concentrations in Central Osaka. *Atmospheric Environment*, 32, pp. 2007-2016.

Müller, J.-F. (1992). Geographical distribution and seasonal variation of surface emissions and deposition velocities of atmospheric trace gases. *Journal of Geophysical Research*, 97, pp. 3787-3804.

Muniz, I. and Galindo, A. (2005). Urban form and the ecological footprint of commuting. The case of Barcelona. *Ecological Economics* 55, pp. 799-514.

National Research Council (1982). *Diesel cars: benefits, risks and public policy: final report of the Diesel Impacts Study Committee/Washington*, DC: National Academy Press.

Newman, P. and Kenworthy, J.R. (1989). Gasoline consumption and cities: a comparison of US cities with a global survey, *Journal of the American Planning Association*.

Pilidis, G.A., Karakitsios, S.P. and Kassomenos, P.A. (2005). BTX measurements in a medium-sized European city. *Atmospheric Environment*, 39, pp. 6051-6065.

Radwan, C. (1995). Particulate emissions trends predicted in the PART5 model. *Air & Waste Management Association*, pp. 599-607.

Rentz, O., Remmers, J., Plinke, E. (eds.) (1987). *Proceedings of the Workshop on Emission Control costs, 28.09.-01.10.1987, Esslingen am Neckar, Germany. Executive Body for the Convention on Long-range Transboundary Air Pollution.* Institute for Industrial Production (IIP) University of Karlsruhe, Karlsruhe, Germany.

Rudolf, W. (1994). Concentration of air pollutants inside cars driving on highways and in downtown areas. *The science of the total environment*, 146/147, pp. 433-444.

Sahu, L.K. and Lal, S. (2006). Distribution of C2-C5 NMHCs and related trace gases at a tropical urban site in India. *Atmospheric Environment*, 40, pp. 880-891.

Schauer, J.J., Kleeman, M.J., Cass, G.R. and Simoneit, B.R.T. (2002). Measurement of emissions from air pollution sources. 5. C1-C32 organic compounds from gasolinepowered motor vehicles. *Environmental Science and Technology*, 36, pp. 1169-1180.

Solberg, S., Dye, C. and Schmidbauer, N. (1996). Carbonyls and non-methane hydrocarbons at rural European sites from Mediterranean to the Arctic. *Journal of Atmospheric Chemistry*, 25, pp. 33-66.

Solberg, S., Dye, C., Schmidbauer, N., Wallasch, N. and Junek, R. (2002). *VOC measurements 2000*. EMEP/CCC-Report 8/2002. http://www.nilu.no/pro jects/ccc/reports/cccr8-2002.pdf.

Srivastava, A., Joseph, A.E. and Devotta, S. (2006). Volatile organic compounds in ambient air of Mumbai – India. *Atmospheric Environment*, 40, pp. 892-903.

Tengstrom, E. and Thynell, M. (1997). *Towards Sustainable Mobility: transporting people and goods in the Baltic Region*, The Baltic University Programme.

U.S. EPA (2005). *The original list of hazardous air pollutants*. http://www.epa.gov/ ttn/atw/pollsour.html

Van Wijnen, J.H. at al. (1995). The exposure of cyclists, car drivers and pedestrians to traffic-related air pollutants. *International archives of occupational and environmental health*, 67, pp. 187-193.

Wackernagel, M. and Rees, W. (1996). *Our ecological footprint. Reducing human impact on the earth*, New Society Publisher, Gabriola Island.

Watson, J.G., Chow, J.C., Fujita, E.M. (2001). Review of volatile organic compound source apportionment by chemical mass balance. *Atmospheric Environment*, 35, pp. 1567-1584.

Wegener, M. (1995). Current and future land use models, *Land Use Modelling Conference Proceedings*, Arlington (TX).

Whitby, K.T. and Cantrell, B. (1976). *Fine particles. International Conference on Environmental Sensing and assessment*, Las Vegas, NV, Institute of Electrical and Electronic Engineers.

III. Monitoring and modelling urban environmental quality

The assessment of transport-induced urban environmental quality covers the whole process from the collection of raw data and the storage and retrieval of this data for computation/modelling, to the presentation of information on environmental quality in terms of occurrence, distribution and concentration. Usually the process is known as "Monitoring, Modelling, Mapping", and supports sustainable decisions for urban environmental management. Generally speaking, monitoring data is used in modelling systems, which provide spatially attributed information to traffic planners and decision-makers. Mapping in most cases is used to visualise processed data in order to derive thematic maps.

3.1 Monitoring ambient air quality

Ambient air sampling presents two critical challenges. Firstly, concentrations of most atmospheric pollutants in ambient air are very low. Analysis of low-level organic compounds in the air requires large sample volumes. Secondly, large variations in analyte concentrations can occur within a short period of time due to meteorological effects. Ambient air samples should be collected at a location that most closely approximates the actual human exposure. The choice of sample location is extremely important for accuracy (Zhang, 2007). The key is to find a place that is either representative of the local air quality, or, depending on objectives, representing a worst case. Documentation of sampling conditions, including meteorological (wind direction, wind speed, temperature, atmospheric stability, atmospheric pressure, humidity, and precipitation), and topographical (buildings, mountains, hills, valleys, lakes, and sea) data, will help later data interpretations.

As indicated in the EU Community Framework Directive on Ambient Air Quality (96/62/EC), the related daughter directives establish limit values intended to avoid, prevent, or reduce harmful effects on human health and the environment as a whole. *Limit value* shall mean a level fixed on the basis of scientific knowledge; it should be attained within a given period and not be exceeded once attained. The first daughter directive (1999/30EC) defined limit values for the concentration in ambient air of sulphur dioxide, nitrogen dioxide and oxides of nitrogen, particulate matter, and lead. Directive 2000/69/EC established limit values for carbon monoxide and benzene. As requested by directive 96/62/EC, limit values must also be fixed for ozone (under negotiation in the EC), poly-aromatic hydrocarbons, cadmium, arsenic, nickel, and mercury. The limit values laid down in the above directives are minimum requirements, allowing member states the possibility of introducing more stringent protective measures and stricter limit values.

The Directive of the European Parliament and of the Council (2008/50/EC) on Ambient Air Quality and Cleaner Air for Europe determines upper and lower assessment thresholds for sulphur dioxide, nitrogen dioxide and oxides of nitrogen, particulate matter (PM10 and PM2.5), lead, benzene and carbon monoxide in ambient air within a zone or agglomeration. Consistency of ambient air monitoring data should meet data quality objectives established by the Directive.

Several key definitions applied to the Directive 2008/50/EC can be found below.

Zone shall mean part of the territory of a Member State, as delimited by that Member State for the purposes of air quality assessment and management.

Agglomeration shall mean a zone that is a conurbation with a population in excess of 250 000 inhabitants, or where the population is 250 000 or less, with a given population density per km^2 to be established by the Member States.

Upper assessment threshold shall mean a level below which a combination of fixed measurements and modelling techniques and/or indicative measurements may be used to assess ambient air quality.

Lower assessment threshold shall mean a level below which modelling or objective-estimation techniques alone may be used to assess ambient air quality. The EU threshold limits for sulphur dioxide, nitrogen dioxide, oxides of nitrogen, particulate matter (PM10/PM2.5), lead, benzene and carbon monoxide are presented in the tables below.

Table 2: Threshold limits for sulphur dioxide

	Health protection	**Vegetation protection**
Upper assessment threshold	60% of 24-hour limit value (75 μg/m^3, not to be exceeded more than 3 times in any calendar year)	60% of winter critical level (12 μg/m^3)
Lower assessment threshold	40% of 24-hour limit value (50 μg/m^3, not to be exceeded more than 3 times in any calendar year)	60% of winter critical level (12 μg/m^3)

Table 3: Threshold limits for nitrogen dioxide and oxides of nitrogen

	Hourly limit value for the protection of human health (NO_2)	**Annual limit value for the protection of human health (NO_2)**	**Annual critical level for the protection of vegetation and natural ecosystems (NO_x)**
Upper assessment threshold	70% of limit value (140 µg/m³, not to be exceeded more than 18 times in any calendar year)	80% of limit value (32 µg/m³)	80% of critical level (24 µg/m³)
Lower assessment threshold	50% of limit value (100 µg/m³, not to be exceeded more than 18 times in any calendar year)	65% of limit value (26 µg/m³)	65% of critical level (19.5 µg/m³)

Table 4: Threshold limits for particulate matter (PM10/PM2.5)

	24-hour average PM10	**Annual average PM10**	**Annual average PM2,5 (*)**
Upper assessment threshold	70% of limit value (35 µg/m³, not to be exceeded more than 35 times in any calendar year)	70% of limit value (28 µg/m³)	70% of limit value (17 µg/m³)
Lower assessment threshold	50% of limit value (25 µg/m³, not to be exceeded more than 35 times in any calendar year)	50% of limit value (20 µg/m³)	50% of limit value (12 µg/m³)
(*) The upper assessment threshold and the lower assessment threshold for PM2.5 do not apply to the measurements to assess compliance with the PM2.5 exposure reduction target for the protection of human health.			

Table 5: Threshold limits of lead

	Annual average
Upper assessment threshold	70% of limit value (0.35 µg/m³)
Lower assessment threshold	50% of limit value (0.25 µg/m³)

Table 6: Threshold limits benzene

	Annual average
Upper assessment threshold	70% of limit value (3.5 µg/m³)
Lower assessment threshold	40% of limit value (2 µg/m³)

Table 7: Threshold limits carbon monoxide

	Eight-hour average
Upper assessment threshold	70% of limit value (7 mg/m^3)
Lower assessment threshold	50% of limit value (5 mg/m^3)

Exceedance of upper and lower assessment thresholds shall be determined on the basis of concentrations during the previous five years, where sufficient data is available. An assessment threshold shall be deemed to have been exceeded if it has been exceeded during at least three separate years out of those previous five years. Where fewer than five years' data is available, member states may determine excesses by combining measurement campaigns of short duration, carried out during the period of the year and at locations likely to be typical of the highest pollution levels, with results obtained from emission inventories and modelling.

In many metropolitan areas, air quality and air pollution is monitored on a regular basis. The data collected is used in a number of contexts. The general goal is to protect human health and the environment. When developing new roads, housing and industries, for example, reliable data and predictions based on this data will provide important background for making the right decisions. Among the users of such data are municipalities, counties, companies and authorities. Faulty decisions will affect human health; projects may even have to be cancelled. In the Stockholm area, the early cost of air pollution is estimated to be SEK 3 billion (1999). The air pollution monitoring system is thus a congenial investment in terms of air quality management.

Errors in environmental data acquisition can be minimized through proper design and the implementation of a quality programme. A quality programme is generally split into two main parts: quality control (QC) and quality assurance (QA). QC is generally a system of *technical* activities aimed at controlling the quality of data so that it meets the requirements of of the user. QC procedures should be specified to measure the accuracy and bias of the data. A QA program is a *management* system that ensures the QC is working as intended.

Fig. 5: Management of Kaunas air quality monitoring system

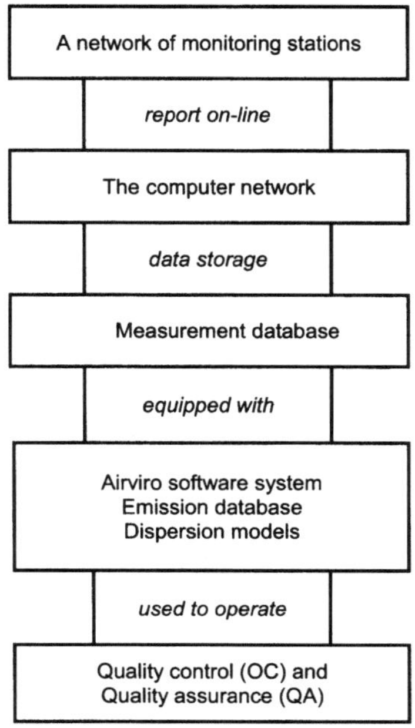

As an example, we can look at the monitoring system operating in Kaunas, Lithuania (Kliucininkas L., 2003a). Air quality monitoring in Kaunas started in the early 70s, and was based on manual sampling and, thereafter, on laboratory analysis using wet chemistry methods. Usually data was available in quarterly reports that were often out-of-date and not regular. Existing databases did not reflect current air pollution trends, and did not meet the needs of city planners and managers.

As the result of the inadequate performance of the Kaunas air quality monitoring system as supervised by the Ministry of Environment, Kaunas Municipality started a parallel air quality monitoring system. In 1993, the *Municipal Programme for Air Quality Monitoring in Kaunas* was approved by the Ministry of Environment, and air quality monitoring was started.

In 1998, the Department of Environment Protection at Kaunas Municipality initiated the PHARE project, and was granted financial support for the *Establishment of Automated Air Quality Monitoring System in Kaunas.*

At present, the Municipal Air Quality Monitoring System in Kaunas comprises three stationary air quality monitoring stations and a mobile laboratory. Sites for monitoring stations were selected according to the national urban air quality monitoring requirements and EU recommendations. One monitoring station is situated in the city centre (Centras), a second is in Šilainiai, the densely populated residential district, and the third is in Dainava, an area with high traffic intensity. The measurement properties are: CO, SO_2, NO, NO_2, NO_X, BTEXs, O_3, particulate matter concentrations and meteorological parameters. The characteristic distribution of NO_2 concentrations during the October-December period are presented in Figures 6 and 7.

Fig. 6: Distribution of NO_2 concentrations on a 24-h basis

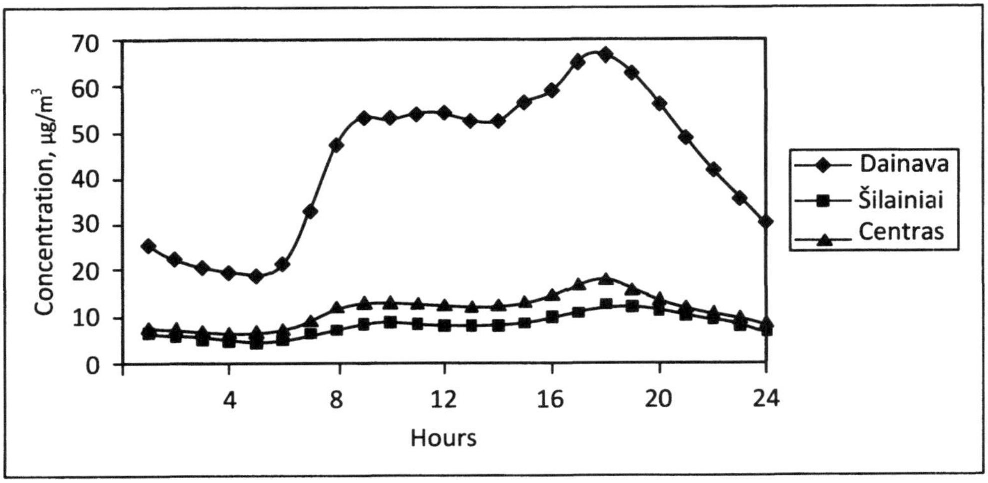

Fig. 7: Distribution of NO_2 concentrations on weekly basis

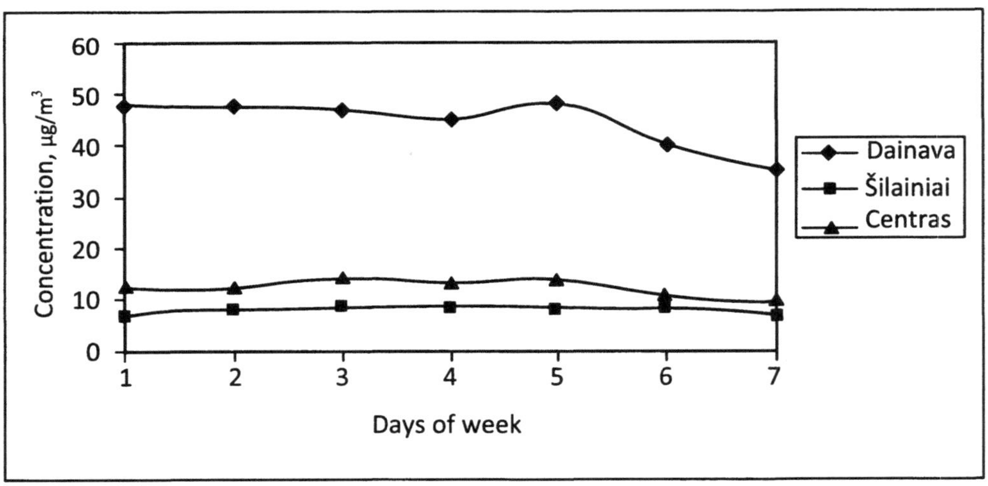

Air quality and meteorological data are collected automatically every hour from the network of stationary monitoring stations and reported to the system via a telecommunication system. The instruments sample the ambient air at roof level in the three monitoring stations. The street air quality in particular is monitored at one street-crossing in Kaunas with an average traffic flow of more than 30,000 vehicles per day. Real time data is transferred to the main computer and analysed with AIRVIRO software. The AIRVIRO software enables dispersion modelling and allows the development of air pollution maps. In addition to automatic analysers, complementary techniques such as passive samplers are used.

In 1999, Kaunas City Council established the municipal public enterprise *Kauno miesto aplinkos kokybės tyrimai*, responsible for checking the environmental quality in the city. The main activities of the enterprise cover studies of air quality and traffic noise, as well as recommendations for improvements to the environmental situation in the city.

The Department of Environment Protection at Kaunas Municipality developed the *Municipal Ambient Air Quality Monitoring and Management Programme*, which in early 2000 was approved by the Lithuanian Ministry of Environment. In the Programme, Kaunas municipal ambient air quality monitoring and management is considered an integral part of environmental monitoring. It covers local air quality monitoring in the administrative boundaries of Kaunas city, and ensures measurements meet the automated air quality monitoring requirements established by the Lithuanian Ministry of Environment.

Among the several methods used to measure air pollution, the following four are most often used for outdoor air quality.

Chemilluminescence for NO. The air is mixed with ozone in a reaction vessel. Nitrogen monoxide in the sample is converted to nitrogen dioxide in an activated state, which will emit light as it returns to its original state. A photomultiplier tube measures the emitted light, which is proportional to the amount of nitrogen monoxide in the sample. Measurements may be taken every five minutes, but normally data is collected each hour.

DOAS, Differential Optical Absorption Spectroscopy. Monochromatic light sweeping over a wide wavelength region is passed through a sample of air, and the absorption spectrum is registered in comparison to a reference beam. Observing the position and height of absorption peaks allows the quantification of NO, NO_2, ozone and some hydrocarbons. Measurements may be taken every five minutes, but normally data are collected every hour.

Active filter methods. The air is passed (pumped) through a filter or a cell with an absorption material which specifically adsorbs a defined component, such as NO_2 or SO_2. Particles of a size above 10μm (PM10) or 2.5μm (PM2.5), as well as black smoke (soot), are also measured with this method. Measurements are made over a period of 24 hours or longer. The filters are collected and the adsorbed substance eluted and measured in the laboratory.

Passive diffusion methods. A measuring cell contains specific adsorbents for the substance to be analysed. Air is passively diffused into the cell. The method is mainly used for NO, NO_2, SO_2, CO, O_3 and VOC. Measurements are taken over a longer period, often many days or weeks. The filters are collected and the adsorbed substance eluted and measured in the laboratory.

Box 1

Passive Sampling for Transport Induced BTEX Monitoring

Objective – to assess concentrations of BTEX (benzene, toluene, ethylbenzene, xylenes) and determine their relationships with the intensity of urban traffic flows.

Methods. The study consisted of two passive sampling campaigns, which took place in Kaunas, Lithuania (pop. 361,930; total area 157 km^2). Measurement campaigns lasting one week were carried out in July and December 2005. Axial Perkin Elmer diffusive samplers with pre-packed non-porous sorbent Carbopack X were used to estimate concentrations of BTEX (benzene, toluene, xylene, ethylbenzene, n-hexane) in ambient air. During each sampling campaign, up to 20 simultaneous locations were used to sample concentrations of BTEX. Samplers were located in close vicinity to the street and were distributed over the entire city area. The instrumental analysis of samples involved the use of an automated thermal desorption (ATD) injector connected to a gas chromatographer (GC) with a flame ionization detector (FID). The sample was thermally desorbed from the Carbopack X onto a cold trap, which then was rapidly heated. The sample was then injected into a GC column. Laboratory and field tests were performed according to the Swedish standard method SS-EN 13528-2:2002 (E) and SS-EN 13528-1. For quality assurance purposes, in several locations duplicate and triplicate samplers were placed in the same location.

Traffic analysis involved measurements of flow intensity, composition, and vehicle speed. It was concluded that the weekly "commuting circle" fairly well reflects temporal variations of traffic flows, and is more or less stable over the year. The analysis revealed that, based on 24-hour time series, the average traffic intensity for the central area of Kaunas city varied little over the whole year.

The obtained concentration values were analysed with respect to the traffic, meteorological and distance factors. The analysis of the street traffic flows was based on the manual counts and automated traffic intensity analysis.

Fig. 8: Relationships between concentration of benzene (upper), toluene (lower) and actual intensity of traffic flows

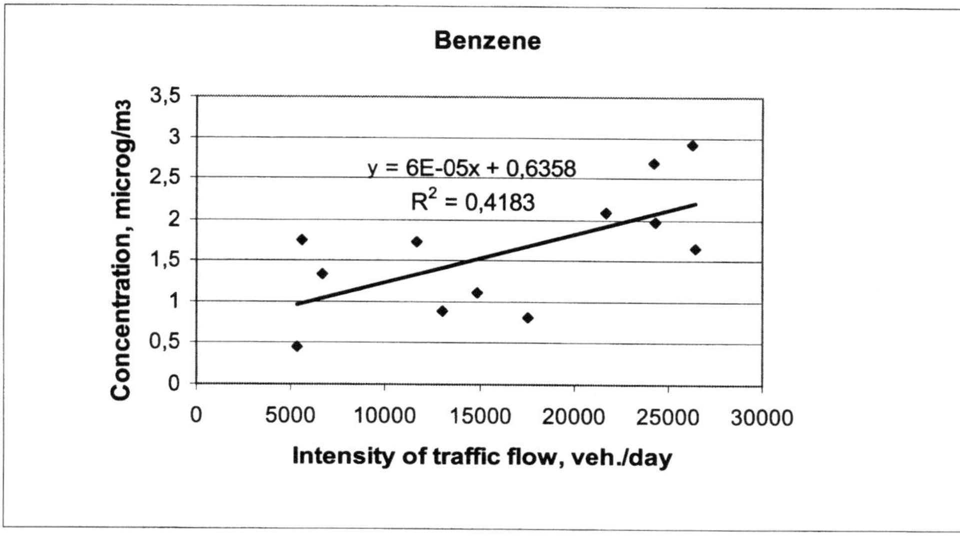

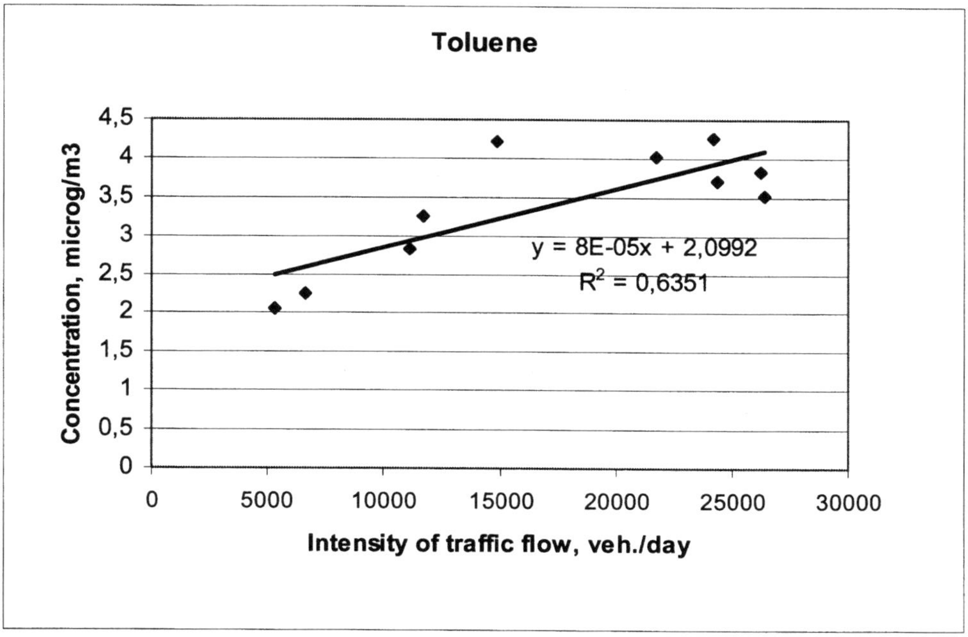

Conclusions

- An analysis of experimental data showed seasonal variations of BTEX concentrations. The concentration values indicated during the December sampling campaign were on average 3-4 times higher than those registered during July. Higher concentrations during the winter campaign could be explained by unfavourable dispersion conditions, slower chemical transformations, and the "cold start time" of longer vehicles.
- An estimation of BTEX concentrations at increasing distance from the street revealed that the concentration drop of benzene and toluene was constant but not substantial; in the case of etylbenzene and xylene, the concentration drop was rapid, and at 50 m distance from the street these compounds were not registered.
- Results of the study provide prerequisites for applying passive diffusive sampling methodology to monitor transport-induced BTEXs. Concentrations of BTEXs from a widespread net of simultaneous sampling locations, combined with the actual traffic flows, could serve as a planning tool to help optimize urban traffic flows and decrease expenditures allocated for environmental monitoring.

Source: Mockuviene, J., Martuzevicius, D., Bergqvist, P.-A., Hoang, K., Obudulu, O., Strandberg, B., Kliucininkas, L. (2006). Passive sampling for transport induced BTEX monitoring. 2nd International Passive Sampling Workshop and Symposium, (IPSW). Bratislava, Slovakia, 3rd-6th May, 2006, p. 63, ISBN 80-7080-602-8.

3.2 Air quality indicators

As a consequence of the increase in air pollution, especially in large urban areas, many international organisations and governments have adopted a series of regulations to keep pollutant concentrations at a low level. Over several years, thorough scientific research has been carried out in order to find out to what extent humans are affected by pollutants diffused in the air. In the course of this research the environment has been monitored chemically and physically, and biological indicators have also been used to evaluate the impact pollutants have on the ecosystem and on humans (Rapport et al., 2003). For instance, several studies have established that nitrogen dioxide (NO_2) and ozone (O_3) increase the risk of death in patients with severe asthma (Sunyer et al., 2002). Ozone polluting the air generates increases in lung cancer rates (Yang et al., 2005). Ozone and carbon monoxide (CO) are linked to cardiac birth deficiencies (Ritz at al., 2002). Traffic-related air pollution increases mortality (Hoek et al., 2003).

In order to prevent the pollutant concentration from being dangerous to health, in several areas the responsible authorities sometimes have to take extraordinary measures, such as imposing limited or total restrictions on traffic circulation. Public awareness of this problem has been heightened considerably in the past few years thanks to media campaigns. A high level of air pollution is not only harmful and annoying to the population, but is also a heavy drain on the wealth of a country.

For these reasons, several countries have introduced evaluation methods that rapidly and efficiently indicate the air quality conditions for the population. First of all, the US Environmental Protection Agency (EPA) developed the air quality index (AQI) (United States Environmental Protection Agency, 1999). This indicator provides air quality information, using numeric and chromatic indicators in order to report immediately and extensively the level of risk for human health.

In all the metropolitan statistical areas (MSAs) of the US with a population of more than 350,000, the AQI provided a report on the safety limits of the following five air pollutants, according to safety limits outlined in the Clean Air Act: ozone (O_3), particulate matter (PM), carbon monoxide (CO), sulphur dioxide (SO_2), nitrogen dioxide (NO_2) (US EPA, 1999).

In each area, the concentrations of these five pollutants is measured for the previous 24 hours (or estimated) and reported according to six reference categories (see Table 8).

Table 8: The six reference categories of the five pollutants embraced by the AQI

O_3, 8-hour mean value (µg/m³)	**PM_{10}, 24-hour mean value (µg/m³)**	**CO, 8-hour mean value (µg/m³)**	**SO_2, 24-hour mean value (µg/m³)**	**NO_2, 1-hour mean value (µg/m³)**	**AQI reference values**
0-137	0-54	0-5.5	0-97	a	0-50
138-180	55-154	5.6-11.76	98-412	a	51-100
181-223	155-254	11.77-15.5	413-640	a	101-150
224-266	255-354	15.6-19.25	641-869	a	151-200
267-800	355-424	19.26-38.0	870-1,727	1,330-2,542	201-300
> 800	425-604	38.1-50.5	1,728-2,300	2,543-4,182	301-500

A CE regulations do not set a nitrogen dioxide (NO2) short-term limit.

According to the Clean Air Act, an AQI value is 100 or less for an acceptable concentration of each pollutant. Therefore, AQI values lower than 100 are judged admissible. For higher AQI values, the air is unhealthy for the most sensitive subjects, and as the air pollution increases it becomes unhealthy also for the general population. It is interesting to notice that the above thresholds are considerably higher than the European Union (EU) regulations.

For each area, the daily AQI value is given by the worst registered condition among the five pollutants:

$$AQI = \max\left\{I_{O_3}, I_{PM_{10}}, I_{CO}, I_{SO2}, I_{NO2}\right\} \tag{1}$$

The sub-index for each pollutant is calculated using the following equation (linear interpolation):

$$I_p = I_{L,p} + \frac{I_{H,p} - I_{L,p}}{BP_{H,p} - BP_{L,p}}\left(C_p - BP_{L,p}\right) \tag{2}$$

where I_p is the sub-index for the *p*th pollutant; C_p is the concentration of the *P*th pollutant; $BP_{H,p}$ is the breakpoint that is greater than or equal to C_p; $BP_{l,p}$ is the breakpoint that is less than or equal to C_p; $I_{H,p}$ is the AQI value corresponding to $BP_{H,p}$; $I_{L,p}$ is the AQI value corresponding to $BP_{L,p}$.

The EPA qualitative descriptions related to the AQI categories are as follows:

1. "Good". The AQI value is within the 0-50 range. The air quality is satisfactory, with very little risk to the population.
2. "Moderate". AQI is between 51 and 100. The air quality is admissible; however, the health of some people could be damaged. For instance, ozone-sensitive people may experience respiratory symptoms.
3. "Unhealthy for sensitive groups". Children who are active outdoors and adults with respiratory problems are at risk due to the ozone exposure; most at risk, however, are people with cardiovascular disease, due to the carbon monoxide exposure. When the AQI value is between 101 and 150, these vulnerable individuals could experience worsened symptoms to the point where their health is compromised. However, much of the population is not at risk in these conditions.
4. Within the 151-200 range, the AQI is considered to be "unhealthy". There is the potential for illness amongst the general population. The most vulnerable individuals could be damaged more seriously.
5. "Very unhealthy". AQI values – between 201 and 300 – represent an alarm. The health of the whole population could be seriously damaged.
6. "Hazardous" AQI values – over 300 – trigger an immediate alarm. There could be severe consequences for the health of the whole population.

European countries have consequently adopted similar systems. This section illustrates the Atmo index, formulated and made operational by the French Environment Ministry (Ministère de l'écologie et du développement durable, 2004), and the *Indice di Qualita dell'Aria* (IQA) index, developed and implemented in some northern regions of Italy: Piedmont, Lombardy. These indicators are similar in pattern, but are operated in different ways. In brief, they take into account the concentrations of the main air pollutants and set a tolerability scale for each one, trying to gather general information on the overall air condition (Franceschini F. et al., 2005).

The Atmo index is based on the concentration of four air pollutants: ozone (O_3), particulate matter (PM_{10}), suphur dioxide (SO_2), and nitrogen dioxide (NO_2). Each of the pollutants is related to a sub-index. The Atmo value is the maximum of the four sub-indexes:

$$Atmo = \max\{I_{O3}, P_{PM10}, I_{SO2}, I_{NO2}\} \quad (3)$$

Each pollutant concentration is measured and reported on a ten-level scale. The first level corresponds to an excellent air quality; the fifth and sixth levels are just around the European "long-term" norms; the eighth level corresponds to the EU "short-term" norms; and finally the tenth level corresponds to conditions hazardous to health. The PM_{10} value is the average of the daily mean values, registered in the different operative stations of the monitoring network. The value of the other pollutants is the mean of the maximum hourly values, registered in the different operative stations of the monitored area. The maximal Atmo index value is 8. In this condition the air quality is very unhealthy.

In some northern Italian regions, different systems are currently being tested to monitor the air quality and inform the public. The analysis will focus on the IQA index, which took inspiration from the US EPA AQI, but incorporated some clear differences.

The IQA aggregates the most critical air pollutants at each time of the year on the basis of their effects on human health: ozone (O_3) and particulate matter (PM_{10}) and nitrogen dioxide (NO_2) in wintertime.

The IQA is expressed by a numerical index between 1 and 7; the higher the air pollution, the more serious the health hazard and index respectively.

The IQA index comprises several sub-indexes, each of which is related to a monitored pollutant. The IQA value is the arithmetic mean of the maximum values of the sub-indexes:

$$IQA = \frac{I_1 + I_2}{2} \quad (4)$$

where I_1 and I_2 are two sub-indexes (among three pollutants) with the higher value.

The IQA is calculated daily using the concentrations of the pollutants over the previous 24 hours. A numerical IQA value equal to 100 essentially corresponds to the air quality safety level for polluting substances. IQA values less than 100 are generally satisfactory, with no potential hazard for public health.

The more the IQA value is over 100, the more the air quality is considered unhealthy – initially only for the most sensitive groups and thereafter for the rest of the population. In short, IQA is a conventional indicator used to:

- report the quality of the air on a daily basis;
- identify the worst environmental parameters;
- calculate the risk that the population is exposed to.

What the three systems that have been discussed have in common that each aggregative indicator "has no memory" for the most critical pollutant. Given a particular combination of air pollutants, each of the three indicators leads to a different model. Each indicator maps the pollutants' real concentrations onto a different scale (Finkelstein, 2003). Another significant aspect of the three indicators is that the environmental conditions they are connected to can be very different. In other words, comparable overall indexes do not entail comparable physical concentrations of the air pollutants.

References

Council directive 1999/30/EC of 22 April 1999 relating to limit values for sulphur dioxide, nitrogen dioxide and oxides of nitrogen, particulate matter and lead in ambient ait. *Official Journal of the European Union*, L. 163/41, 29.6.1999.

Council directive 96/62/EC of 27 September 1996 on ambient air quality assessment and management. *Official Journal of the European Union,* L 296, 21.11.1996.

Directive 2000/69/EC of the European Parliament and of the Council of 16 November 2000 relating to limit values for benzene and carbon monoxide in ambient air. *Official Journal of the European Union* L313/12, 13.12.2000.

Directive 2008/50/EC of the European Parliament and of the Council of 21 May 2008 on ambient air quality and cleaner air for Europe. *Official Journal of the European Union* L152/1, 11.6.2008.

Finkelstein, L. (2003). Widely, strongly and weakly defined measurement, *Measurement*, vol. 34 no. 1, pp. 39-48.

Franceshini, F., Galetto, M. and Maisano, D. (2005). A short survey on air quality indicators: properties, use, and (mis)use. *International Journal Management of Environmental Quality,* vol. 16 no. 5, pp. 490-504.

Hoek, G., Brunekreef, B., Goldholm, S., D. van den Brandt, P.A. (2003). Cardiopulmonasry mortality and air pollution, *The lancet*, vol. 363 no. 9384, p. 587.

Kliucininkas, L. (2003a). Municipal ambient air quality monitoring – experience from Kaunas. In the book *Environmental Science* (eds. Ryden, L., Migula, P., Andersson, M.). The Baltic University Press. ISBN: 91-970017-0-8, pp. 490-491.

– (2003b). Local Sustainability Indicators. *Urban Environment Management.* Baltic University Press, Uppsala, vol. 4, pp. 13-26.

Mockuviene, J., Martuzevicius, D., Bergqvist, P.-A., Hoang, K., Obudulu, O., Strandberg, B. and Kliucininkas, L. (2006). *Passive sampling for transport induced BTEX monitoring.* 2nd International Passive Sampling Workshop and Symposium, (IPSW). Bratislava, Slovakia, 3rd-6th May, p. 63, ISBN 80-7080-602-8.

Rapport, D.J., Howard, J., Lannigan, R. and McCauley, W. (2003). Linking health and ecology in the medical curriculum, *Environment International,* vol. 29 no. 2/3, pp. 353-358.

Ritz, B., Yu, F., Fruin, S., Chapa, G., Shaw, G.M. and Harris, J.A. (2002). Ambient air pollution and risk of birth defects in Southern California, *American Journal of Epidemiology*, vol. 155 no. 1, pp. 17-25.

Sunyer, J., Baqsagana, X., Belmonte, J. and Anto, J.M. (2002). Effect of nitrogen dioxide and ozone on the risk of dyingin pateinsts with severe asthma, *Thorax (International Journal of respiratory Medicine),* vol. 57 no. 8, pp. 687-693.

United States Environmental Protection Agency (1999). *Guideline for reporting of daily air quality – air quality index*, EPA-454/R-99-010, EPA, North Carolina.

Yang, I.A., Holz, O., Jorres R.A., Magnussen, H., Barton, S.J., Rodriguez, S., Cakebread, J.A., Holloway, J.W. and Holgate, S.T. (2005). Association of tumor necrosis factor – polymorphisms and ozone-induced change in lung function, *American Journal of Repiratory and Critical Care Medicine*, vol. 171, pp. 171-176.

Zhang, C.C. (2007). *Fundamentals of environmental sampling and analysis.* John Wiley & Sons.

3.3 Case study I. Indoor and outdoor concentrations of fine particles, particle-bound PAHs and volatile organic compounds in Kaunas, Lithuania

***Source**: Kliucininkas, L., Martuzevicius, D., Krugly, E., Prasauskas, T., Kauneliene, V., Molnar, P., Strandberg, B. (2011). Indoor and outdoor concentrations of fine particles, particle-bound PAHs and volatile organic compounds in Kaunas, Lithuania. // Journal of Environmental Monitoring. Cambridge: The Royal Society of Chemistry. ISSN 1464-0325, vol. 13, iss. 1, pp. 182-191.*

This complex study presents indoor and outdoor levels of airborne fine particles, particle-bound PAHs and VOCs at two urban locations in the city of Kaunas, Lithuania, and considers possible sources of pollution. Two sampling campaigns were carried out in January-February and March-April 2009. The mean outdoor $PM_{2.5}$ concentration at Location 1 in winter was 34.5±15.2 μg/m^3, while in spring it was 24.7±12.2 μg/m^3; at Location 2 the corresponding values were 36.7±21.7 and 22.4±19.4 μg/m^3, respectively. In general there was little difference between the PM concentrations at Locations 1 and 2. $PM_{2.5}$ concentrations were lower during the spring sampling campaign. These PM concentrations were similar to those in many other European cities; however, the levels of most of the PAHs analysed were notably higher. The mean sum PAH concentrations at Locations 1 and 2 in the winter campaign were 75.1±32.7 and 32.7±11.8 ng/m^3, respectively. These differences are greater than expected from the difference in traffic intensity at the two sites, suggesting that there is another significant source of PAH emissions at Location 1 in addition to the traffic. The low indoor/outdoor (I/O) ratios that were observed indicate that PAH emissions at the locations studied arise primarily from outdoor sources. The buildings at both locations have old windows with wooden frames that are fairly permissive in terms of air circulation. VOC concentrations were mostly low and comparable to those reported from Sweden. The mean outdoor concentrations of VOCs were: 0.7±0.2, 3.0±0.8, 0.5±0.2, 3.5±0.3, and 0.2±0.1 μg/m^3, for benzene, toluene, ethylbenzene, sum of m-, p-, o-xylenes, and naphthalene, respectively. Higher concentrations of VOCs were observed during the winter campaign, possibly due to slower dispersion, slower chemical transformations and/or the lengthy "cold start time" required by vehicles in wintertime. A trajectory analysis showed that air masses coming from eastern Europe carried significantly higher levels of $PM_{2.5}$ compared to masses from other regions, but the PAHs within the $PM_{2.5}$ are of local origin. It has been suggested that street dust, widely used for winter sanding activities in Eastern and Central European countries, may act not only as a source of PM, but also as source of particle-bound PAHs. Other potential sources include vehicle exhaust, domestic heating and long-range transport.

Introduction

Particulate matter (PM) in the ambient air has been the subject of numerous studies and PM levels have been routinely monitored for many years. Recently, attention has focused on fine particles ($PM_{2.5}$) due to their deep penetration into the respiratory system. The association of fine particles with mutagenic and carcinogenic compounds such as polycyclic aromatic hydrocarbons (PAHs) may contribute to acute health effects and potentially result in long-term health risks (Li et al., 2009). PAHs are products of incomplete combustion and are often generated through open burning, incineration, industrial power generation, and vehicle emissions. Among other pollution sources, surface depositions on roads/highways contain many toxic micropollutants such as heavy metals and PAHs (Barbosa et al., 1999; Berbee et al., 1999). Typical outdoor concentrations of benzo(a)pyrene, as reported in Duisburg, Amsterdam and Helsinki, range from ca 0.1 to 1 ng/m^3.[4] However, notably higher BaP outdoor concentrations have been reported in Prague and Zagreb (3.0 and 3.2 ng/m^3, respectively) (Saarnio et al., 2008; Sisovic et al., 2005). Diverse PAHs, with varying properties and origins, may be adsorbed to particular matter, and increasing attention has been paid to the profiles in environmental samples (Li et al., 2009; Saarnio et al., 2008; Guo et al., 2003; Tang et al., 2005; Sracquadanio et al., 2007; Helleén et al., 2008; Rajput et al., 2009). In fact, it has been well-documented that indoor PAH concentrations are influenced by both indoor and outdoor sources (Chuang et al., 1991; Ott et al., 1994).

In recent decades, much of the concern over indoor and outdoor air quality has also focused on volatile organic compounds (VOCs), such as benzene, ethylbenzene, toluene, and xylenes (BETXs) (Hansen and Palmgren, 1996; Cocheo et al., 2000; Ilgen et al., 2001; Bertoni et al., 2002). This is because VOCs can readily enter the body via inhalation. Consequently, long-term exposure to VOCs presents a substantial health risk, even at low concentrations (Buczynska et al., 2009). In economically developed countries the main outdoor source of VOCs is road traffic. For instance, studies performed in Denmark in 1997/1998 indicated that benzene emissions from traffic strongly contribute not only to urban concentrations, but also to exposure and domestic concentrations.[18] This study reports outdoor urban concentrations for benzene, toluene, and the sum of xylenes of 9.6, 25.7, and 27.6 μg/m^3, and indoor air concentrations of 3.5, 25.3, and 9.4 μg/m^3, respectively. Similarly, a study on air pollution in Germany concluded that, in street canyons, high-density traffic is the primary source of BETX compounds in indoor air (Ilgen et al., 2001a; Ilgen et al., 2001b).

Despite the concern regarding VOCs, data on their levels and distribution in Lithuania are sparse, and to our knowledge, limited to information collected in two outdoor measurement campaigns (Kliucininkas). Thus, the aim of this study was to investigate indoor and outdoor levels of PM, particle-bound PAHs and VOCs at two urban locations in Kaunas, Lithuania. In addition to presenting the results, the study discusses possible sources of these pollutants, such as road traffic, aerosolized street dust and long-range transport.

Methods

Sampling locations and experimental set-up

Two sampling locations in the city of Kaunas (pop. 361,930; total area 157 km^2), Lithuania, were chosen. Both locations are urban and may be classified as "traffic" sampling sites, since they are located in close proximity to streets with dense traffic. The first location (henceforth referred to as Location 1) was beside a two-lane street (Vyduno al.) with an average daily traffic of over 12 000 vehicles, with pronounced peaks during rush hours. This street connects the city centre to outer residential areas and carries a moderate level of bus traffic: approximately 300 diesel-fuelled public buses and 850 diesel-fuelled microbuses (important public transport vehicles in Kaunas) travel along it every day.

The second location (Location 2) was on a one-way three-lane street (Kestucio g.). This street is in the city centre, with an average daily traffic of approx. 19 000 vehicles, including 450 diesel public buses and 1 820 microbuses. The traffic peaks were less pronounced in this location.

In Location 1 air was sampled at an indoor site inside a room and an outdoor site on the balcony on the 3rd floor of a 5-story university dormitory, located 12 m away from the street. The room was occupied by two non-smoking students who carried out their routine activities (no thermal aerosol generation was expected). Outdoor air was sampled at Location 2 from a storage facility on the first floor of a university faculty building, 3 m away from the street. The indoor site was an unoccupied room visited only occasionally by sampling crew and other university staff. The sampling locations are depicted in Fig. 1.

In order to estimate seasonal fluctuations in the atmospheric abundance of the two main PM size fractions ($PM_{2.5}$ and PM_{10}), and the particle-bound PAH content of these fractions, daily samples were taken at the outdoor sites over two-week periods in January and April of 2009 at Location 1, and in February and March of 2009 at Location 2. In addition, the levels of the PM_4 fraction and VOCs at both indoor and outdoor sites were monitored in week-long daily sampling campaigns; at Location 1 samples were obtained in January and April of 2009, while at Location 2 samples were taken in February and March of 2009. In each case samples were collected (as described below) over eight-hour periods during weekdays.

Fig. 1: Schematic sketches of sampling locations

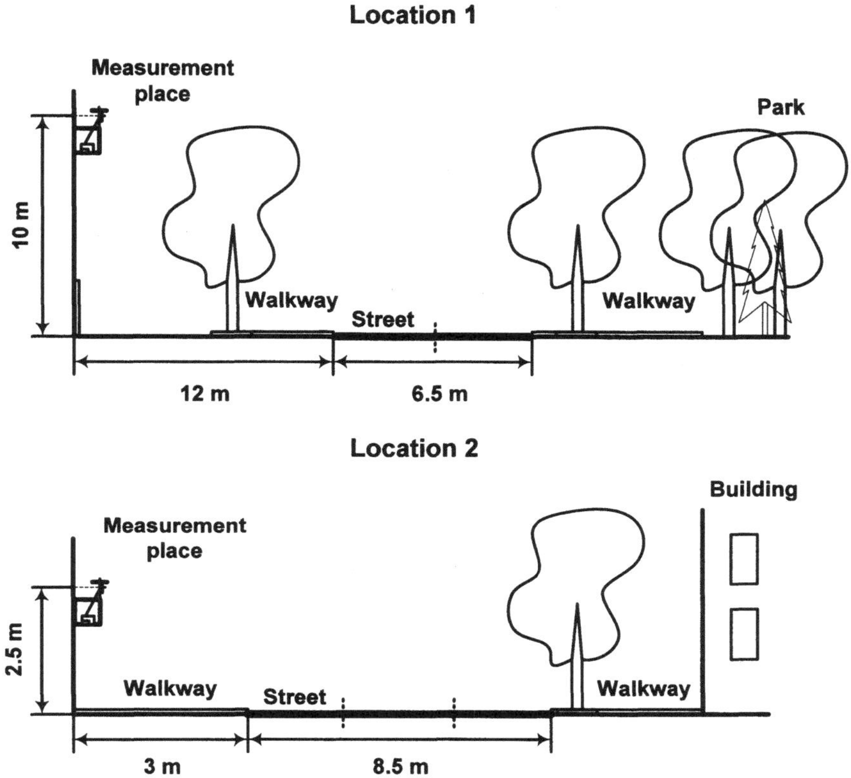

PM Sampling and concentration measurement

Airborne particulate matter was collected on glass microfibre filters (GF/A, Whatman International Ltd., Maidstone, UK; diameter 25 mm, pore size 1.6 µm) over a sampling period of 8 hours. Before use, the microfibre filters were heated in an oven for 4 hours at 500 °C and conditioned for 24 hours at 20±1°C before and after sampling. Finally, PM mass concentrations (µg/m^3) were obtained from the gravimetric analysis of samples, using a MXA5 microbalance (Radwag, Poland).

The PM_{10} and $PM_{2.5}$ fractions were separated at an air sampling flow rate of 16.7l min^{-1} (1 m^3/hr) by cyclones (URG Corp, Chappel Hill, NC, USA) with appropriate cut-off sizes, and separate rotary vane vacuum pumps equipped with flowmeters. To minimise the annoyance of indoor noise for residents, the PM_4 fraction was sampled using a flow rate of 2.2 l min^{-1} (0.13 m^3/hr) by SKC Conductive Plastic Cyclones for respirable dust sampling with personal sampling pumps.

Table 1: *Names, abbreviations, numbers of rings and vapour pressures of the PAHs analysed*

Name	Abbreviation	Number of rings	Vapour Pressure, Pa[c]
naphthalene	Nap	2	8.64
2-methylnaphthalene	2MeN	2	7.156
1-methylnaphthalene	1MeN	2	9.07
biphenyl	Bph	2	0.78
2,3-dimethylnaphthalene	3MeBPh	2	1.55
2,3,5-trimethylnaphthalene	3MeNap	2	N/A
acenaphthylene	Acy	3	0.893
acenaphthene	Ace	3	0.161
fluorene	Fl	3	0.0425
1-methylfluorene	1MeFl	3	0.032
phenanthrene	Phe	3	0.881
1-methylphenanthrene[a]	1MePhe	3	0.0186
2-methylphenanthrene	2MePhe	3	N/A
3-methylphenanthrene	3MePhe	3	N/A
anthracene	Ant	3	0.00102
1-methylanthracene	1MeAnt	3	N/A
2-phenylnaphthalene	2PheNap	3	N/A
fluoranthene[a]	Fla	4	0.000707
1-methylfluoranthene	1MeFla	4	N/A
pyrene	Pyr	4	0.000335
1-methylpyrene	MePyr	4	N/A
retene	Ret	4	N/A
benzo(a)anthracene	BaA	4	0.0000266
chrysene[a]	Chry	4	0.19
2-methylchrysene	2MeCry	4	N/A
benzo(b)fluoranthene[a,b]	BbF	5	0.0000667
benzo(k)fluoranthene[a,b]	BkF	5	0.000067
benzo(a)pyrene[a,b]	BaP	5	N/A
perylene[a]	Per	5	N/A
dibenzo(a,h)anthracene[a,b]	DbA	5	0.000000133
indeno(1,2,3-c,d)pyrene[a,b]	IdP	6	0.000000133
benzo(g,h,i)perylene[a]	BghiP	6	0.0000000133

a: genotoxic, WHO/IPCS (1998)
b: possibly carcinogenic, WHO/IPCS (1998)
c: Handbook of Physical-Chemical Properties and Environmental Fate for Organic Chemicals (2006)

PAH analysis

After gravimetric analysis the filters were shipped to the Sahlgrenska Academy at the University of Gothenburg, Sweden, for PAH analysis. Deposited particles were extracted from the filters by means of a Sonica ultrasonic extractor (Soltec Srl, Italy) in 3 mL of dichloromethane for 10 min. Before extraction, 40 μL of an internal standard mixture containing the 16 deuterated US EPA priority PAHs, at a concentration of 1.0 ng μL^{-1}, was added to the samples. The extract solutions were filtered through Pasteur pipettes filled with Na_2SO_4, then concentrated by purging with ultra-pure nitrogen gas to 200-300 μL. Prior to analysis, a recovery estimation standard (octachloronaphthalene) was added and the excess solvent was evaporated to the final volume of 20-30 μL. PAHs were then separated and detected by means of high-resolution gas chromatography/low resolution mass spectrometry (HRGC/LRMS). The MS instrument (a 5973 model connected to a 6890N GC, both made by Agilent Technologies, Inc., Santa Clara, CA, USA) was operated in electron impact (EI) ionization, selected ion monitoring (SIM) mode. The GC column was a non-polar capillary column (60 m x 0.32 mm i.d. and 0.25 μm film thickness; J&W DB-5, Folsom, USA). The SIM descriptors included the most abundant ion of each native compound and the ^2H-labeled PAH standards. In total, the concentrations of 32 PAHs were determined (Table 1). It should be noted that, although a relatively low flow rate was used in this study, the levels of the more volatile PAHs (those with 2-4 rings) may have been somewhat underestimated due to losses from the particles during sampling.

VOC sampling and analysis

VOCs were actively sampled using Perkin Elmer (PE) tube samplers containing 300 mg Tenax TA and a flow rate of 9.9 ± 0.4 mL/min. After sampling, the time was noted and the samplers were retrieved, sealed and shipped to the laboratory at Sahlgrenska Academy at the University of Gothenburg, Sweden, for instrumental analysis.

The samples were analysed using a Unity Markes International Limited thermal desorber (Unity Ultra TD) connected to a gas chromatograph (6890, Agilent Technologis, Inc., Santa Clara, CA, USA), with a selective mass detector (5973, Agilent Technologies, Inc.). Each sample tube was thermally desorbed for 5 min at 320°C, focused on a cold trap at – 10°C and desorbed at a maximum temperature of 320°C for GC injection. The GC oven temperature was maintained at 50°C for 3 min, then raised at 10°C/min to 180°C, and finally at 10°C/min to 300°C, at which temperature it was held for 5 min. The VOCs (benzene, toluene, ethylbenzene, xylenes and naphthalene) were detected using SIM, including the two most abundant ions of each compound. Controls for the quantification and identification of target compounds were established by injecting standard solutions (5 µL) of all target compounds in methanol into Perkin Elmer tubes in a stream of helium (15 mL min^{-1}).

Quality control

For the PAH analyses a certified reference material (SRM 1649a urban dust) was used as a quality control (QC) sample. The measured levels of the 12 PAHs rarely deviated more than 10 per cent from the certified levels. For the VOC analyses, QC samples obtained from the National Physical Laboratory (NPL), Teddington, UK, were analysed at the same time as the samples. The levels of the VOCs detected in the QC samples did not deviate more than 5 per cent from the certified levels. The QC results were considered acceptable during the whole study. Blanks were processed in parallel with the samples in this study. Some PAH residues were found on blank filters and all VOC target compounds were found in the PE blanks, but in no case was the amount of any given compound in any blank greater than 10 per cent of the amount found in any non-blank sample.

Modelling long-distance transported pollution

Air mass back trajectories were computed using the NOAA ARL HYSPLIT Model to investigate the effect of long-distance transported (LDT) pollution (Draxler and Rolph, 2003). For each sampling day, 96-hour air mass back trajectories were computed at start-up time and stop time at three starting heights (100, 200, and 500 m above ground level). The trajectories were either divided into four trajectory classes or sectors, representing different source areas – Eastern Europe, Western Europe, the Nordic region (Finland, Norway, Sweden and the Baltic countries) and an eastern sector given the name “Russia” (including Russia, Belarus and Ukraine) – or remained undetermined (for trajectories that shifted classes during the sampling day). These sectors are similar to those used in a recent publication from Estonia (Orru et al., 2010). The classification was based on the criterion that all trajectories during a sampling period must share a major path belonging to the same class.

Statistical calculations were performed using Wilcoxon’s rank sum test implemented in the SAS System for Windows, version 9.2 (SAS, 2008). Statistical significance refers to $p<0.05$ in two-tailed tests.

Table 2: Comparison of $PM_{2.5}$ and PM_{10} concentrations ($\mu g/m^3$) in world cities and Locations 1 and 2 (this study), and at the automated monitoring station, EPA

	Kaunas, Lithuania													
	Location 1	**Location 2**	**Automated monitoring station, EPA**	Oxford, Ohio	Madrid, Spain	Bern, Switzerland	Athens, Greece	Belgrade, Serbia	Beirut, Liban	Teheran, Iran	Istambul, Turkey	Barcelona, Spain	Berlin, Germany	San Paulo, Brazil
$PM_{2.5}$[a]	34.5 ± 15.2	36.7 ± 21.7	26.81 ± 12.43	15.7	34	24.6	40.2	61	40	24.3	27.7	27.7	30	20.9
$PM_{2.5}$[b]	24.7 ± 12.2	22.4 ± 19.4	23.19 ± 11.75											
PM_{10}[a]	35.2 ± 14.6	39.7± 20.3	33.91 ± 20.03	16.3	48	40.2	75.5	72	76	122.1	40.4	40.6	38	10.7
PM_{10}[b]	41.7 ± 11.8	34.2 ± 20.6	32.55 ± 22.88											
	Current study			Wojas B et al., 2007	Artiñano B et al., 2003	Hueglin C et al., 2005	Chaloulakou A et al. 2003[27]	Tasic M.D. et al., 2005	Shaka H et al., 2004	Kakooei H et al., 2007	Karaca F et al., 2005	Querol X, et al., 2001	Lenschow P, et al., 2001	Miranda R et al., 2007
a – winter sampling period														
b – spring sampling period														

Results and discussion

PM

The $PM_{2.5}$ and PM_{10} outdoor concentrations measured in this study are compared with those reported in various other cities around the world in Table 1. The results from the winter campaign (January-February) revealed very minor differences in $PM_{2.5}$ and PM_{10} concentrations (the average $PM_{2.5}/PM_{10}$ ratio was 0.95 at Location 1 and 0.90 at Location 2). This indicates that both fractions were generated by the same source. Most of the PM_{10} mass consisted of fine particles, which probably originated from street dust. The average concentrations of particulate matter in Kaunas were comparable to those in most other European sites. In general there was little difference between the PM concentrations at Locations 1 and 2. $PM_{2.5}$ concentrations were lower during the spring sampling campaign; for PM_{10} no significant seasonal concentration variations were observed. The average recorded $PM_{2.5}$ and PM_{10} concentrations were compared to those obtained from the Lithuania Environmental Protection Agency's automated monitoring station at Kaunas. The average recorded concentrations were comparable to EPA monitored concentrations (see Table 2). At Location 1, the winter $PM_{2.5}$ concentration was 34.5±15.2 $\mu g/m^3$ while in spring it was 24.7±12.2 $\mu g/m^3$; at Location 2 the corresponding values were 36.7±21.7 and 22.4±19.4 $\mu g/m^3$, respectively.

PAHs

Very little information on the abundance and distribution of PAHs in Lithuania is available. Previously published studies are limited to investigations of BaP concentrations at a background station located on the Baltic sea coast and levels of 16 PAHs in the air (using passive sampling on PUF type samplers and soil at five sites in the country) (Milukaite, 2006; Milukaite et al., 2008).

The outdoor PAH concentrations measured at the two sampling locations in this study are presented in Fig. 2. High molecular weight PAHs with five to six rings, such as benzo(a)anthracene, chrysene, benzo(a)pyrene, benzo(g,h,i)perylene predominated, while low molecular weight compounds were present only in smaller quantities. High concentrations of the 4-ring PAHs fluoranthene and pyrene were also observed. These findings are as expected: 90 per cent of 5-6 ring PAHs are adsorbed on particles and only 10 per cent partition into the vapour phase; 4-ring PAHs are found in similar amounts in both phases, while PAHs with 2-3 rings are found predominantly (>90%) in the vapour phase (Odabasi et al., 1999). Five PAH components, i.e. acenaphthylene, 2,3,5-trimethylnaphthalene, 1-methylfluoranthene, 1-methylfluorene, and 2-methylchrysene, were not detected in any samples.

The mean sum PAH concentrations at Locations 1 and 2 in the winter campaign were 75.1±32.7 and 32.7±11.8 ng/m^3, respectively. This difference is greater than expected from the difference in traffic intensity at the two sites, suggesting that there is another significant source of PAH emissions at Location 1 in addition to the traffic.

Fig. 2: Mean concentrations of PAHs measured in the PM2.5 fraction in the winter sampling period

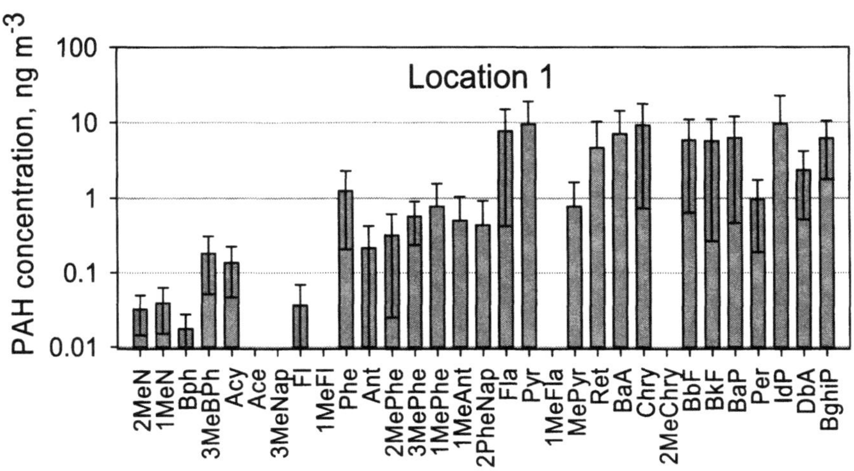

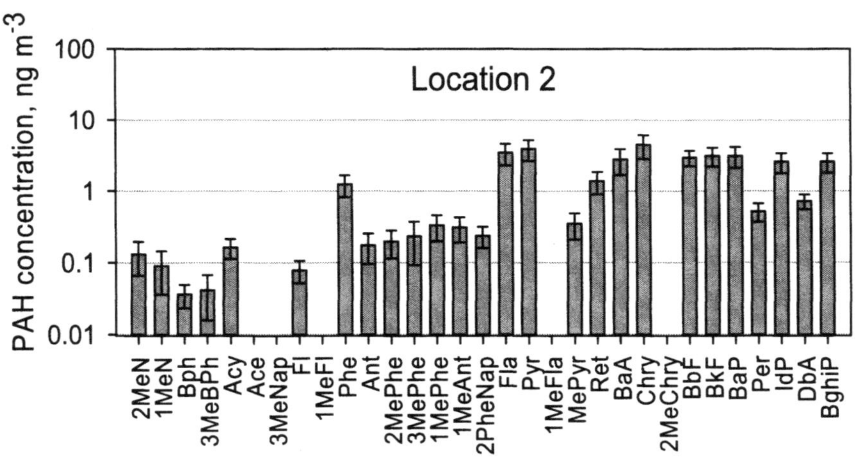

As shown in Table 3, the concentrations of most PAHs observed in this study are higher than those reported by other authors, particularly at Location 1. Comparison of benzo(a)pyrene levels in the $PM_{2.5}$ fraction reported by Saarnio et al. (2008) showed that in Western European cities concentrations varied from 0.1 to 1.1 ng/m^3, and in Eastern and Central European cities from 3.0 to 3.2 ng/m^3. Benzo(a)pyrene levels at Location 2 were similar to those reported in Prague and Zagreb. It should be noted that the European limit value for benzo(a)pyrene is 1 ng/m^3 in PM_{10}.

Table 3: Comparison of average PAH concentrations measured in the $PM_{2.5}$ fraction, ng/m^3, in this study (Locations 1 and 2) and in other cities

Compound	This study Location 1	This study Location 2	Kurkimäki Finland Hellen et al. 2008	Bologna Italy Stracquandanio et al. 2007	Duisburg Germany Saarnio et al. 2008	Prague Czech R. Saarnio et al. 2008	Amsterdam Netherlands Saarnio et al. 2008	Helsinki Finland Saarnio et al. 2008	Zagreb Croatia Sisovic et al. 2005	Atlanta USA Li et al. 2009
	winter	winter	winter	winter	winter	winter	winter	spring	winter	winter
phenanthrene	1.3±0.6	1.2±0.4	2.5	0.9	0.7	4.8	0.8	0.3		0.15
anthracene	0.2±0.1	0.2±0.1	0.4							0.02
fluoranthene	7.6±3.4	3.5±1.2	4.4		1.6	6.4	1.3	0.5	3.7	0.14
pyrene	9.5±4.1	3.9±1.3	4.0	1.4	1.3	5.6	0.9	0.4	4.7	0.17
retene	4.6±1.3	1.4±0.5								0.14
benzo(a)anthracene	7.0±3.0	2.8±1.1	1.3	1.1						0.19
chrysene	9.2±4.1	4.5±1.7	1.6	2.2						0.23
benzo(b)fluoranthene	5.8±2.0	3.0±0.7		2.5					3.5	0.61
benzo(k)fluoranthene	5.6±1.9	3.1±0.9		0.9					2.1	0.18
benzo(a)pyrene	6.2±3.9	3.2±1.0	1.3	1.7	1.1	3.0	0.3	0.1	3.2	0.42
indeno(1,2,3-c,d)pyrene	9.6±1.8	2.6±0.8	2.2							0.59
dibenzo(a,h)anthracene	2.3±1.9	0.7±0.2	0.4							0.02
benzo(g,h,i)perylene	6.1±2.9	2.6±0.8	1.5	2.7					4.1	0.31
Sum PAHs	75.1±32.7	32.7±11.8								3.17

Table 4: Spearman rank correlation matrix between PAH compounds

Location 1 (above the diagonal; row labels on the right) / **Location 2** (below the diagonal; row labels on the left)

	Nap	2MeNap	1MeNap	Bhp	Acy	Fl	Phe	Ant	2MePhe	1MePhe	1MeAnt	2PhNap	Fla	Pyr	MePyr	Ret	BaAnt	Chry	BbF	BkF	BaP	Per	IdP	DbA	BghiP	
Nap		0.61	0.72	0.68	0.71	0.75	0.84	0.76	0.76	0.72	0.74	0.75	0.66	0.67	0.66	0.64	0.55	0.59	0.63	0.63	0.58	0.40	0.54	-0.05	0.34	Nap
2MeNap	0.82		0.62	0.48	0.73	0.67	0.58	0.45	0.58	0.65	0.58	0.58	0.52	0.52	0.42	0.49	0.30	0.32	0.42	0.37	0.24	0.14	0.25	0.09	-0.02	2MeNap
1MeNap	0.83	0.97		0.79	0.34	0.59	0.46	0.13	0.30	0.25	0.27	0.28	0.14	0.14	0.11	0.12	-0.04	-0.03	0.08	0.06	0.00	-0.22	0.20	-0.41	-0.21	1MeNap
Bhp	0.53	0.77	0.82		0.60	0.78	0.59	0.46	0.49	0.45	0.49	0.51	0.38	0.37	0.35	0.35	0.22	0.27	0.34	0.31	0.22	-0.07	0.26	-0.35	-0.16	Bhp
Acy	0.77	0.90	0.93	0.72		0.92	0.92	0.88	0.85	0.82	0.80	0.77	0.71	0.70	0.66	0.69	0.60	0.64	0.65	0.63	0.57	0.45	0.37	0.21	0.38	Acy
Fl	0.75	0.92	0.97	0.73	0.93		0.87	0.75	0.68	0.59	0.61	0.58	0.49	0.45	0.41	0.39	0.35	0.41	0.35	0.37	0.37	0.21	0.07	0.17	0.13	Fl
Phe	0.75	0.92	0.95	0.73	0.98	0.97		0.95	0.95	0.91	0.93	0.92	0.87	0.85	0.84	0.82	0.79	0.80	0.81	0.82	0.77	0.65	0.65	0.15	0.60	Phe
Ant	0.75	0.92	0.95	0.73	0.98	0.97	1.00		0.98	0.97	0.97	0.96	0.94	0.93	0.92	0.89	0.87	0.90	0.89	0.90	0.81	0.71	0.64	0.20	0.65	Ant
2MePhe	0.75	0.92	0.95	0.73	0.98	0.97	1.00	1.00		0.98	0.99	0.98	0.96	0.95	0.93	0.91	0.91	0.91	0.92	0.93	0.87	0.78	0.70	0.21	0.70	2MePhe
1MePhe	0.72	0.88	0.93	0.68	0.97	0.98	0.98	0.98	0.98		0.99	0.98	0.98	0.97	0.96	0.95	0.91	0.92	0.95	0.95	0.84	0.78	0.74	0.23	0.73	1MePhe
1MeAnt	0.78	0.95	0.98	0.78	0.95	0.98	0.98	0.98	0.98	0.97		0.99	0.97	0.96	0.95	0.93	0.92	0.92	0.94	0.95	0.87	0.79	0.73	0.18	0.71	1MeAnt
2PhNap	0.75	0.92	0.95	0.73	0.98	0.97	1.00	1.00	1.00	0.98	0.98		0.98	0.97	0.96	0.93	0.91	0.92	0.93	0.95	0.86	0.76	0.77	0.13	0.70	2PhNap
Fla	0.75	0.92	0.95	0.73	0.98	0.97	1.00	1.00	1.00	0.98	0.98	1.00		0.99	0.99	0.97	0.95	0.96	0.97	0.98	0.85	0.84	0.83	0.23	0.81	Fla
Pyr	0.72	0.88	0.93	0.68	0.97	0.98	0.98	0.98	0.98	1.00	0.97	0.98	0.98		0.99	0.98	0.96	0.97	0.98	0.99	0.86	0.84	0.84	0.24	0.81	Pyr
MePyr	0.77	0.88	0.93	0.67	0.90	0.98	0.95	0.95	0.95	0.97	0.97	0.95	0.95	0.97		0.98	0.97	0.97	0.99	0.99	0.87	0.86	0.85	0.23	0.83	MePyr
Ret	0.75	0.92	0.95	0.73	0.98	0.97	1.00	1.00	1.00	0.98	0.98	1.00	1.00	0.98	0.95		0.93	0.93	0.97	0.97	0.84	0.82	0.83	0.20	0.78	Ret
BaAnt	0.73	0.85	0.90	0.62	0.93	0.97	0.97	0.97	0.97	0.98	0.95	0.97	0.97	0.98	0.98	0.97		0.99	0.97	0.98	0.93	0.95	0.74	0.38	0.88	BaAnt
Chry	0.73	0.53	0.58	0.33	0.62	0.63	0.63	0.63	0.63	0.65	0.62	0.63	0.63	0.65	0.73	0.63	0.75		0.97	0.98	0.92	0.93	0.75	0.38	0.87	Chry
BbF	0.72	0.83	0.87	0.53	0.90	0.95	0.93	0.93	0.93	0.97	0.92	0.93	0.93	0.97	0.97	0.93	0.98	0.73		0.98	0.88	0.88	0.82	0.24	0.84	BbF
BkF	0.73	0.85	0.90	0.62	0.93	0.97	0.97	0.97	0.97	0.98	0.95	0.97	0.97	0.98	0.98	0.97	1.00	0.75	0.98		0.90	0.89	0.83	0.27	0.85	BkF
BaP	0.73	0.53	0.58	0.33	0.62	0.63	0.63	0.63	0.63	0.65	0.62	0.63	0.63	0.65	0.73	0.63	0.75	1.00	0.73	0.75		0.88	0.57	0.25	0.69	BaP
Per	0.63	0.78	0.85	0.63	0.92	0.93	0.95	0.95	0.95	0.97	0.92	0.95	0.95	0.97	0.95	0.95	0.98	0.73	0.95	0.98	0.73		0.60	0.57	0.93	Per
IdP	0.73	0.85	0.90	0.62	0.93	0.97	0.97	0.97	0.97	0.98	0.95	0.97	0.97	0.98	0.98	0.97	1.00	0.75	0.98	1.00	0.75	0.98		-0.06	0.72	IdP
DbA	0.53	0.45	0.47	0.10	0.63	0.57	0.58	0.58	0.58	0.65	0.50	0.58	0.58	0.65	0.58	0.58	0.67	0.67	0.75	0.67	0.67	0.63	0.67		0.59	DbA
BghiP	0.73	0.85	0.90	0.62	0.93	0.97	0.97	0.97	0.97	0.98	0.95	0.97	0.97	0.98	0.98	0.97	1.00	0.75	0.98	1.00	0.75	0.98	1.00	0.67		BghiP
	Nap	2MeNap	1MeNap	Bhp	Acy	Fl	Phe	Ant	2MePhe	1MePhe	1MeAnt	2PhNap	Fla	Pyr	MePyr	Ret	BaAnt	Chry	BbF	BkF	BaP	Per	IdP	DbA	BghiP	

Location 2

The abundance of compounds such as benzo(a)anthracene, chrysene, benzo(a)pyrene, and benzo(g,h,i)perylene, which are characteristic of vehicular emissions, suggests that traffic is probably one of the most important sources of PAH emissions at the investigated sites in Kaunas. However, although the traffic intensity is lower at Location 1 than at Location 2, levels of certain individual PAHs at Location 1 were found to be up to four times greater than at Location 2, possibly because most buildings are heated by the district supply system at Location 2, while at Location 1 most households have independent heating systems. To identify the sources of the various PAH emissions more clearly, the data were subjected to correlation analysis, factor analysis and a comparative analysis of indicative ratios and fractions. The intercorrelation matrix of the measured PAH concentrations is shown in Table 4.

As can be seen from the table, inter-correlation among PAH concentrations are generally very high for most of PAHs analysed. At Location 2 low molecular weight (LMW) compounds correlate with high molecular weight (HMW) compounds better than at Location 1, but the correlations of chrysene, benzo(a)pyrene and dibenzo(a,h)anthracene with other PAH concentrations is weaker. Diesel fuel reportedly contains high concentrations of LMW compounds, but because of pyrosynthesis during combustion in diesel engines the emission rate of HMW PAHs is greater than that of LMW PAHs (Lim et al., 2005; Marr et al., 1999). Therefore, the high correlations of LMW and HMW PAH concentrations may be due to emissions from diesel engines.

It should be noted that correlations of 1-methylphenanthrene with most of the other PAHs, including high molecular weight compounds, are very high. The high concentrations of alkylated PAHs that were observed, relative to unsubstituted PAHs, may indicate the presence of petrogenic sources (Simo et al., 1997; Gustafson et al., 1997). Although the concentrations of methylated compounds observed in this study were lower than those of the corresponding unsubstituted species, the aforementioned correlations between the concentration of 1-methylphenanthrene and those of most of the other PAH analytes suggests that significant quantities of the observed PAHs were derived from unburned fuel.

Pyrene and chrysene are known to be emitted during industrial oil burning; fluoranthene and pyrene are typically emitted by both petrol and diesel vehicles; chrysene, benzo(b)fluoranthene and benzo(k)fluoranthene are characteristic of emissions from diesel vehicles (Rajput and Lakhani, 2009). The high correlations between the concentrations of chrysene, benzo(b)fluoranthene, benzo(k)fluoranthene, fluoranthene and pyrene corroborate the importance of traffic as a source of PAH emissions at the investigated sites. However, it should be noted that the correlation of retene with HMW PAHs is also very high. Retene has been suggested as an indicator of biomass burning and has a unique formation mechanism or environmental pathway (Li et al., 2009; McDonald et al., 2000). Wood and peat are very commonly used fuels for heating individual houses in Kaunas.

To obtain further indications of the likely origins of the detected PAHs, sum concentrations of nine major combustion-derived PAHs (CPAHs); fluoranthene, pyrene, chrysene, benzo(b)fluoranthene, benzo(k)fluoranthene, benzo(a)pyrene, benzo[e]pyrene, indeno(1,2,3-cd)pyrene, benzo(g,h,i)perylene, and CPAH/Sum PAH ratios were calculated. These ratios are reported to be 0.51, 0.41, and 0.30, for emissions from automobiles with catalytic converters, other automobiles, and heavy duty diesel trucks, respectively (Rogge et al., 1993). The CPAH : Sum PAH ratio at Location 1 was as high as 0.78, suggesting that traffic is not the only important source of CPAHs in Kaunas.

Table 5: Results of factor analysis of PAH concentrations

	Location 1					Location 2	
	Factor 1	Factor 2	Factor 3	Factor 4		Factor 1	Factor 2
Nap	**0.65**	0.50	0.54	0.07	**Nap**	0.31	**0.85**
2MeNap	**0.79**	-0.12	0.04	0.57	**2MeNap**	0.35	**0.92**
1MeNap	**0.94**	-0.03	0.32	-0.03	**1MeNap**	0.62	**0.76**
Bph	**0.81**	0.18	0.43	0.27	**Bph**	0.65	0.63
Acy	**0.99**	0.12	0.04	0.03	**Acy**	**0.86**	0.49
Fl	**0.99**	0.07	0.04	-0.09	**Fl**	**0.85**	0.52
Phe	**0.98**	0.16	0.11	-0.04	**Phe**	**0.90**	0.41
Ant	**0.94**	0.30	0.16	0.05	**Ant**	**0.90**	0.43
2MePhe	**0.95**	0.26	0.14	0.00	**2MePhe**	**0.90**	0.42
1MePhe	**0.95**	0.28	0.12	0.07	**1MePhe**	**0.91**	0.40
1MeAnt	**0.90**	0.33	0.24	0.11	**1MeAnt**	**0.92**	0.39
2PheNap	**0.93**	0.32	0.18	0.07	**2PheNap**	**0.92**	0.38
Fla	**0.82**	0.53	0.13	0.18	**Fla**	**0.91**	0.41
Pyr	0.64	**0.73**	0.12	0.21	**Pyr**	**0.90**	0.43
MePyr	0.44	**0.82**	0.22	0.29	**MePyr**	**0.88**	0.47
Re	0.17	0.58	0.32	**0.73**	**Re**	**0.92**	0.38
BaA	0.18	**0.97**	0.12	0.07	**BaA**	**0.86**	0.50
Chry	0.28	**0.95**	-0.02	0.14	**Chry**	**0.90**	0.35
BbF	0.45	**0.83**	0.32	0.07	**BbF**	**0.90**	0.44
BkF	0.37	**0.86**	0.33	0.05	**BkF**	**0.89**	0.46
BaP	0.04	**0.90**	0.07	0.00	**BaP**	**0.89**	0.32
Per	-0.17	**0.96**	-0.18	-0.07	**Per**	**0.87**	0.48
IdP	0.42	0.62	0.56	0.01	**IdP**	**0.87**	0.47
DbaA	-0.35	0.04	**-0.91**	-0.09	**DbaA**	**0.92**	0.18
BghiP	0.05	**0.88**	-0.24	-0.09	**BghiP**	**0.85**	0.51

Factor analysis of the recorded PAH concentrations provided further insights into the influence of traffic and fuel combustion on the air quality in Kaunas (Table 5). The analysis revealed two factors influencing PAH concentrations at

Location 2 and four factors at Location 1. It is apparent that the retene and dibenzo(a,h)anthracene observed at Location 1 arise from a different source from that of the other combustion PAHs. By contrast, at Location 2, it is likely that traffic is the major source of PAHs, with a small quantity of the most volatile PAHs arising from a different source.

Fig. 3: Ratios of indoor to outdoor (I/O) PAH concentrations in PM_4 at Locations 1 and 2

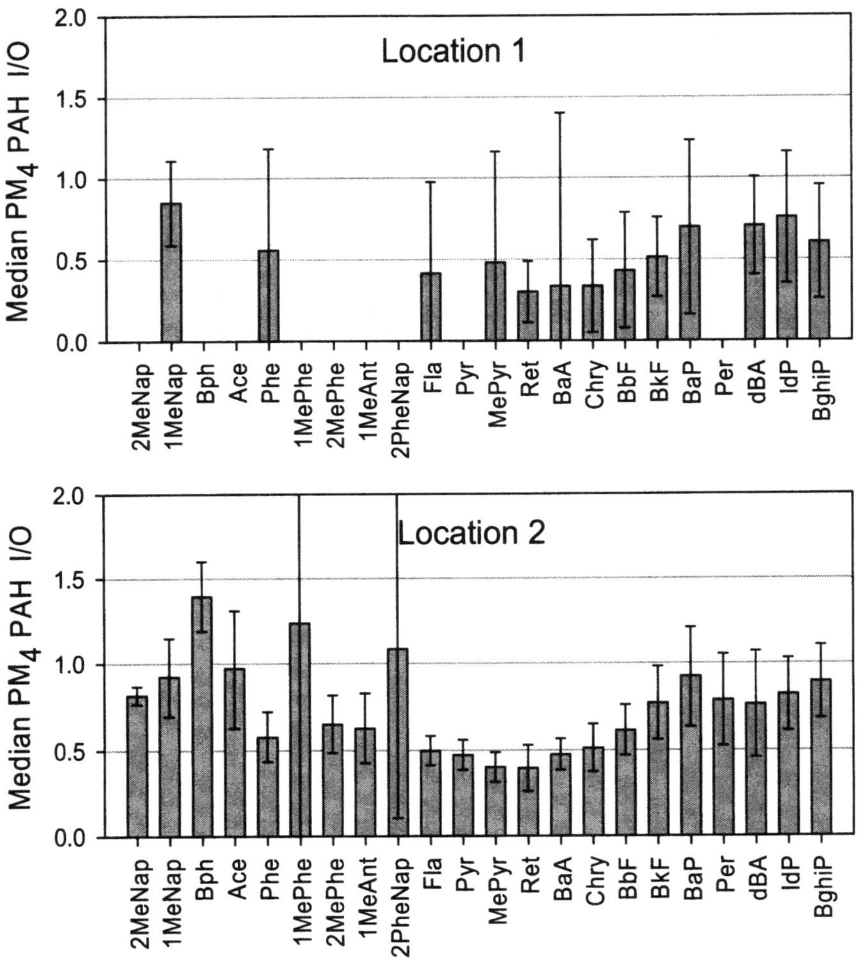

As discussed by Saarnio et al. (2008), neither the levels of individual PAHs nor the ratios of their concentrations can be regarded as highly specific indicators of emission sources, and for several reasons: PAHs originate from a large variety of combustion sources with only slightly different emission profiles; PAHs have widely differing vapour pressures and reactions; and particulate PAH concentrations depend on the ambient temperature and solar radiation intensity as well as on the total particulate mass concentration (Saarnio et al., 2008). Furthermore, some compounds, especially those that are semi-volatile (2-4 rings), may be lost during long sampling campaigns. Nevertheless, their concentration ratios can give hints about the relative importance of possible sources (see Fig. 3).

In addition, indicative PAH concentration fractions and ratios were calculated, and the results were compared to data from a comprehensive study of PAHs in size-segregated particulate matter from European urban sites carried out by Saarnio et al. (2008) and from a similar study by Tang et al. (2005) in China, Japan and Russia (Saarnio et al., 2008; Tang et al., 2005). Only data from cold seasons were included.

The observed indeno(1,2,3-cd)pyrene: (indeno(1,2,3-cd)pyrene + benzo(g,h,i) perylene) ratio is comparable to that reported in other studies, confirming that traffic is a significant source of PAH emissions at both locations examined. This conclusion is further supported by the benzo(a)anthracene:(benzo(a)anthracene + chrysene) and benzo(a)pyrene: (benzo(a)pyrene + chrysene) ratios. The influence of fossil fuel combustion on PAH levels in Kaunas is indicated by the benzo(a)pyrene : benzo(g,h,i)perylene and benzo(a)pyrene : chrysene ratios.

The ratio of airborne PAH concentrations in indoor environments to those in outdoor environments (I/O ratios) can be used to determine whether the major PAH source(s) are located indoors (in which case I/O>1) or outdoors (in which case I/O<1). I/O ratios for PAHs bound to the PM_4 fraction are presented in Fig. 3. The low I/O ratios that were observed may indicate that PAH emissions at the locations studied arise primarily from outdoor sources. It should be noted that the buildings at both locations have old windows with wooden frames that are fairly permissive in terms of air circulation. Thus, residents of older buildings that are close to roads may be significantly exposed to PAHs and PM in general.

VOCs

In developed and industrialized countries, traffic is the most important source of high levels of VOCs in towns, and thus also of the high exposure of the population in the ambient environment. The outdoor and indoor average concentrations, standard deviations, minimal and maximal values of BETX (Benzene, Ethylbenzene, Toluene and Xylene) and Naphthalene observed in the present study are presented in Table 6. The observed outdoor VOC concentrations are lower than those reported in most previous studies. Monitoring of the outdoor air directly outside the windows revealed that the concentration of benzene ranged from 0.3

to 1.7 μg/m^3. The benzene concentration observed during January-February was somewhat higher (by 0.4 μg/m^3) than that observed in March-April. Similar variations in benzene concentrations were observed in both locations. The results obtained for toluene exhibited significant variability, with the recorded concentrations ranging from 0.4 to 9.5 μg/m^3. The average outdoor concentration of toluene at Location 2 was 40 per cent higher than that observed at Location 1. For comparison, Fernandez-Villarrenaga et al. (2004) carried out measurements in A Coruña , a medium-sized town in Spain, and detected toluene, ethylbenzene, *m* + *p*-xylene and *o*-xylene at concentrations of 3.4, 23.6, 3.3, 5.1 and 2.7 μg/m^3, respectively, while a study conducted in Sweden reported urban indoor and outdoor benzene levels to be in the range of 0.4-4 μg/m^3 (Fernandez-Villarrenaga et al., 2004; Strandberg et al., 2006). The average outdoor ethylbenzene concentrations ranged from 0.2 to 1.3 μg/m^3, with the highest concentration being observed at Location 2 in April. The average outdoor concentration of the sum of all xylenes measured during January-February was 40 per cent higher than that measured in March-April, and the concentrations of individual xylene fractions (*m* + *p*-xylene and *o*-xylene) exhibited similar trends to those of the sum of all xylenes. Average outdoor concentrations of naphthalene were higher during January-February than in March-April, and peaked (0.32 μg/m^3) in mid January.

The indoor concentrations of benzene ranged from 0.3 μg/m^3 to 1.4 μg/m^3, and no significant differences in benzene concentration between samples taken in January-February and March-April were found. As with the outdoor concentrations of toluene (see above), there were wide variations in the indoor toluene concentrations, from 0.5 μg/m^3 to 16.3 μg/m^3. The indoor concentrations of ethylbenzene ranged from 0.1 μg/m^3 to 0.5 μg/m^3. For the sum of xylenes, the average of the values measured during January-February was 40 per cent lower than the March-April average. The average indoor concentrations of naphthalene ranged from 0.1 μg/m^3 to 0.3 μg/m^3, and there were no substantial differences between measurements obtained during the winter and spring periods.

The VOC I/O values at both locations are summarized in Table 7. For benzene, it is apparent that at Location 1 the most significant sources are indoors; the I/O ratios observed in January and April were 1.3 and 2.2, respectively. By contrast, the results from Location 2 indicate that the main sources of benzene are outdoors, with I/O ratios < 1 throughout the study. For toluene, both the I/O values and the correlation coefficients exhibited a high degree of variability. For ethylbenzene and xylenes the concentrations in the indoor air were almost equal to those directly outside the windows, except during the sampling week in April, when evidence of strong indoor sources of all the studied VOCs was detected. Thus, for ethylbenzene and xylenes, the outdoor air determines the indoor air, leading to near-equilibrium between indoor and outdoor air. This conclusion is supported by the high correlation coefficients between indoor and outdoor air concentrations of ethylbenzene and xylenes, as shown in Table 7. The results obtained for toluene, ethylbenzene and xylenes correspond with the findings of German authors (Ilgen, E. et al., 2001b). The I/O ratios and correlation coefficients for naphthalene exhibited a high degree of variability.

Table 6: Outdoor and indoor air concentrations of VOCs (μg/m³)

		Measurement week							
		1[a]		2[b]		3[c]		4[d]	
		Outdoor	Indoor	Outdoor	Indoor	Outdoor	Indoor	Outdoor	Indoor
Benzene	**Mean**	**0.9**	**0.7**	**1.0**	**0.7**	**0.7**	**0.5**	**0.4**	**0.8**
	Std.dev.	0.5	0.4	0.5	0.3	0.3	0.2	0.2	0.4
	Min	0.4	0.3	0.5	0.4	0.4	0.3	0.3	0.3
	Max	1.7	1.3	1.6	1.3	1.0	0.7	0.6	1.4
Toluene	**Mean**	**2.7**	**5.3**	**2.0**	**1.8**	**1.2**	**0.9**	**3.5**	**7.0**
	Std.dev.	3.2	3.4	1.0	0.4	0.9	0.5	1.6	8.2
	Min	0.8	2.0	0.7	1.3	0.4	0.5	2.1	0.6
	Max	9.5	9.8	3.3	2.2	2.6	1.4	5.3	16.3
Ethylbenzene	**Mean**	**0.2**	**0.1**	**0.3**	**0.3**	**0.2**	**0.2**	**1.3**	**0.3**
	Std.dev.	0.2	0.1	0.1	0.1	0.1	0.1	1.8	0.2
	Min	0.1	0.1	0.1	0.1	0.1	0.1	0.1	0.1
	Max	0.6	0.2	0.4	0.4	0.4	0.3	2.0	0.5
Σ Xylenes	**Mean**	**0.9**	**0.6**	**1.3**	**1.1**	**0.7**	**0.7**	**0.6**	**1.7**
	Std.dev.	0.9	0.1	0.6	0.4	0.6	0.4	0.3	1.3
	Min	0.2	0.4	0.3	0.5	0.3	0.4	0.3	0.1
	Max	2.7	0.7	2.0	1.5	1.8	1.0	1.0	2.6
Naphthalene	**Mean**	**0.2**	**0.2**	**0.2**	**0.1**	**0.1**	**0.1**	**0.1**	**0.1**
	Std.dev.	0.1	0.1	0.1	0.1	0.1	0.1	0.1	0.1
	Min	0.1	0.2	0.1	0.1	0.1	0.1	0.1	0.1
	Max	0.3	0.3	0.3	0.2	0.2	0.2	0.2	0.2

a: Location 1, winter; b: Location 2, winter, c: Location 1, spring, d: Location 2, spring.

Table 7: Concentration ratios of VOCs of the indoor and outdoor air (I/O ratio) and correlation coefficients between indoor and outdoor air

	Measurement week			
	1[a]	2[b]	3[c]	4[d]
*Concentration **ratio, indoor to** outdoor air (I/O)*				
Benzene	1.3	0.8	0.5	2.2
Toluene	3.6	1.1	0.9	1.7
Ethylbenzene	0.9	1.1	1.0	1.7
Σ Xylenes	1.0	1.0	1.0	1.8
Naphtalene	1.8	1.1	2.6	1.3
Correlation coefficients between indoor and outdoor air				
Benzene	0.10	0.20	0.89	0.50
Toluene	- 0.15	0.26	- 0.11	0.56
Ethylbenzene	0.87	0.57	0.68	0.95
Σ Xylenes	0.79	0.62	0.66	0.73
Naphthalene	- 0.88	0.41	0.52	0.66

a: Location 1, winter; b: Location 2, winter; c: Location 1; spring; d: Location, 2 spring.

Influence of long-distance transported pollution on $PM_{2.5}$ and PAH levels

The trajectory analysis showed that air masses coming from Eastern Europe carried significantly higher levels of pollutants than those from other regions; air masses from other regions in Europe all had similar $PM_{2.5}$ levels (see Table 8). Air masses from the Nordic countries usually carry lower pollutant loads, as the median levels suggest, but high concentrations in masses from the Nordic sector on some days probably explains the lack of significant differences in $PM_{2.5}$ levels (Orru et al., 2010; Johanesson et al., 2007).

For a few of the PAHs (not presented in the table: 2-methylnaphthalene, acenaphthylene, fluorene, phenantrene, and anthracene; also naphthalene), significantly higher concentrations were observed on days when the air originated from Eastern Europe rather than from one or more of the other regions. However, in most cases no such differences were observed.

Table 8: Summary of the trajectory analysis, the four different classes and the number of trajectories belonging to each class (in brackets), mean and median $PM_{2.5}$ concentrations and the p-values for the Wilcoxon's rank sum test between pairs of trajectory classes

	$PM_{2.5}$ (µg m^{-3})		**p-value of the rank sum test**		
	Mean	**Median**	**Russia**	**Western Europe**	**Nordic**
Eastern Europe (9)	45.7	42.0	0.0318	0.0119	0.0464
Russia (10)	27.2	23.7		0.9575	0.4716
Western Europe (6)	23.3	24.9			0.8388
Nordic (13)	25.5	15.4			
Unclassified (5)	25.3	16.8			

The higher concentrations of $PM_{2.5}$ on days when the air originated from Eastern Europe was expected because anthropogenic emissions in Eastern Europe are higher than in the other regions. However, no consistent differences in levels of PAHs in air originating from different regions were observed, suggesting that local sources contribute most of the sampled concentrations of the PAHs.

Conclusions

The mean outdoor $PM_{2.5}$ concentration at Location 1 in winter was 34.5±15.2 µg/m^3, while in spring it was 24.7±12.2 µg/m^3; at Location 2 the corresponding values were 36.7±21.7 and 22.4±19.4 µg/m^3, respectively. The mean PM_{10} concentrations at Location 1 in winter and spring were 35.2±14.6 and 41.7±11.8 µg/m^3, respectively, while at Location 2 the corresponding values were 39.7±20.3 and 34.2±20.6 µg/m^3. $PM_{2.5}$ concentrations were lower during the spring sampling campaign, but no significant seasonal variations in PM_{10} concentrations were observed. This study demonstrates that PM levels in Kaunas are comparable to those observed in many other European locations.

The levels of most PAHs analysed in this study, especially at Location 1, are notably higher than those reported elsewhere in Europe. The mean sum PAH concentrations at Locations 1 and 2 in the winter campaign were 75.1±32.7 and 32.7±11.8 ng/m^3, respectively. The experimental results corroborated the hypotheses that 5-6 and 4-ring PAHs originate from automotive emissions, are bound to street dust and dispersed over the area. Thus, street dust may not only emit fugitive dust, but also be a substantial source of particle-bound PAHs. A trajectory analysis showed that long-range transport is also a source of $PM_{2.5}$ in the region, but that the PAHs within the $PM_{2.5}$ are of local origin.

VOC concentrations were mostly low and comparable to those reported from Sweden. The mean outdoor concentrations of the VOCs benzene, toluene, ethylbenzene, sum of m-, p-, o-xylenes, and naphtalene were: 0.7±0.2, 3.0±0.8, 0.5±0.2, 3.5±0.3 and 0.2±0.1 $\mu g/m^3$ $\mu g/m^3$, respectively. The outdoor concentrations of benzene and xylenes measured during the January-February sampling campaigns were up to 40 per cent higher than those observed during the March-April sampling campaign. Indoor concentrations of benzene did not differ significantly between the January-February and March-April campaigns, but the concentration of xylenes was found to be 40 per cent higher in January-February than in March-April.

Acknowledgements

This study was supported by the Lithuanian State Science and Studies Foundation (Grant No. T–103/09), and the Socrates/Erasmus student mobility programme. The authors are grateful to Ms Inga Vaskeviciute for the PAH laboratory analysis, to the Department of Solid Mechanics at Kaunas University of Technology (Head Assoc. Prof Paulius Griskevicius), and to Ms Loreta Kregzdaite for providing access to measurement sites.

References

Artínano, B., Salvador, P., Alonso, D.G., Querol, W. and Alastuey, A. (2003). *Environmental Pollution*, 125, pp. 453-465.

Barbosa, A.E. and Hvitved, J.T. (1999). *Science of Total Environment*, 235, pp. 151-159.

Berbee, R., Rijs, G., de Brouwer, R. and van Velzen, L. (1999). *Water Environment Research*, 71, pp. 183-190.

Bertoni, G., Ciuchini, C., Pasini, A. and Tappa, R. (2002). *J. Environ. Monit.*, 4, pp. 903-909.

Buczynska, A.J., Krata, A., Stranger, M., Locateli Godoi, A.F., Kontozova-Deutsch, V., Bencs, I., Naveau, I., Roekens, E. and Van Grieken, R. (2009). *Atmospheric Environment*, 43, pp. 311-318.

Chaloulakou, A., Kassomenos, P., Spyrellis, N., Demokritou, P. and Koutrakis, P. (2003). *Atmospheric Environment*, 37, pp. 649-660.

Chuang, J.C., Mack, G.A., Kuhlman, M.R. and Wilson, N.K. (1991). *Atmospheric Environment,* 25, pp. 369-380.

Cocheo, V., Sacco, P., Boaretto, C., De Saeger, E., Ballesta, P., Skov, H., Goelen, E., Gonzalez, N. and Caracena, A.B. (2000): *Nature*, 404, pp. 141-142.

Draxler, R.R. and Rolph, G.D. (2003). *HYSPLIT (Hybrid Single-Particle Lagrangian Integrated Trajectory) Model*. NOAA Air Resources Laboratory, Silver Spring MD.

Fernandez-Villarrenaga, V., Lopez-Mahia, P., Muniategui-Lorenzo, S., Prada-Rodriguez, S., Fernandez-Fernandez, E., Tomas, X. (2004). *Science of Total Environment,* 334, pp. 167-176.

Guo, H., Lee, S.C., Ho, K.F., Wang, X.M. and Zou, S.C. (2003). *Atmospheric Environment*, 37, pp. 5307-5317.

Gustafson, K.E. and Dickhut, R.M. (1997). *Environmental Science and Technology*, 31, pp. 140-147.

Hansen, A.B. and Palmgren, F. (1996). *Science of Total Environment,* 189-190, pp. 451-457.

Helleén, H., Hakola, H., Haaparanta, S., Pietarila, H. and Kauhaniemi, M. (2008). *Sci. Total Environ.*, 393, pp. 283-290.

Hueglin, C., Gehrig, R., Baltensperger, U., Gysel, M., Monn, C. and Vonmont, H. (2005). *Atmospheric Environment*, 39, pp. 637-651.

Ilgen, E., Karfich, N., Levsen, K., Angerer, J., Schneider, P., Heinrich, J., Wichmann, H.E., Dunemann, L. and Begerow, J. (2001). *Atmospheric Environment*, 35, pp. 1235-1252.

Ilgen, E., Levsen, K., Angerer, J., Schneider, P., Heinrich, J. and Wichmann, E. (2001). *Atmospheric Environment*, 35, pp. 1265-1279.

Johanesson, S., Gustafson, P., Molnar, P., Barregard, L. and Sallsten, G. (2007). *Journal of Exposure Science and Environmental Epidemiology*, 17, pp. 613-624.

Kakooei, H. and Kakooei, A.A. (2007). *Journal of Applied Sciences*, 7, 3081-3085.

Karaca, F., Alagha, O. and Erturk, F. (2005). *Chemosphere*, 49, pp. 1183-1190.

Kliucininkas, L.: unpublished work.

Lenschow, P., Abraham, H.J., Kutzner, K., Lutz, M., Preuss, J.D. and Reichenbacher, W. (2001). *Atmospheric Environment*, 35, pp. S23-S33.

Li, Z., Porter, E.N., Sjodin, A., Needham, L.L., Lee, S., Russell, A.G. and Mulholland, J.A. (2009). *Atmospheric Environment*, 43, pp. 4187-4193.

Lim, C.H., Ayoko, G.A., Morawska, L., Ristovski, Z.D. and Jayaratne, E.R. (2005). *Atmospheric Environment*, 39, pp. 7836-7848.

Marr, L.C., Kirchstetter, T.W., Hartey, R.A., Miguel, A.H., Hering, S.V. and Hammond, S.K. (1999). *Environmental Science and Technology*, 33, pp. 3091-3099.

McDonald, J.D., Zielinska, B., Fujita, E.M., Sagebiel, J.C., Chow, J.C. and Watson, J.G. (2000). *Environmental Science and Technology*, 34, pp. 2080-2091.
Milukaite, A. (2006). *Atmospheric Environment*, 40, pp. 2046-2057.
Milukaite, A., Klanova, J., Holoubek, I., Rimselyte, I. and Kvietkus, K. (2008). *Lithuanian Journal of Physics*, 48, pp. 357-366.
Miranda, R. and Tornaz, E. (2008). *Atmospheric Research*, 87, pp. 147-157.
Odabasi, M., Vardar, N., Sofuoglu, A., Tasdemir, Y. and Holsen, T.M. (1999). *Science of Total Environment*, 227, pp. 57-67.
Orru, H., Kimmel, V., Kikas, U., Soon, A., Kunzli, N., Schins, R.P.F., Borm, P.J.A. and Forsberg, B. (2010). *Science of Total Environment,* 408, pp. 1515-1522.
Ott, W., Wilson, N., Klepeis, N. and Switzer, P. (1994). Air and Waste Management Association, Pittsburgh, PA.
Querol, X., Alastuey, A., Rodriguez, S., Plana, F., Ruiz, C.R., Cots, N., Massague, G. and Puig, O. (2001). *Atmospheric Environment*, 35, pp. 6407-6419.
Rajput, N., and Lakhani, A. (2009). *Environmental Monitoring and Assessment*, 150, pp. 273-284.
Rogge, W.F., Hildemann, L.M., Mazurek, M.A., Cass, G.R., and Simoneit, B.R.T. (1993). *Environmental Science and Technology*, 27, pp. 636-651.
Saarnio, K., Sillanpaa, M., Hillamo, R., Sandell, E., Pennanen A.S. and Salonen, R.O. (2008). *Atmospheric Environment*, 42, pp. 9087-9097.
SAS, SAS Statistical software version 9.2, SAS Institute Inc., Cary, NC, 2008.
Shaka, H. and Saliba, N.A. (2004). *Atmospheric Environment*, 38, pp. 523-531.
Simo, R., Grimalt, J.O. and Albaiges, J. (1997). *Environmental Science and Technology*, 31, pp. 2697-2700.
Sisovic, A., Vadjic, Z., Sega, K., Beslic, I. and Vadjic, V. (2005). *Environmental Contamination and Toxicology*, 75, pp. 121-126.
Skov, H., Hansen, A.B., Lorenzen, G., Andersen, H.V., Lofstrom P. and Christensen, C.S. (2001). *Atmospheric Environment*, 35, pp. 2463-2471.
Soderstrom, H.S. and Bergqvist, P.A. (2003). *Environmental Science and Technology*, 37, pp. 47-52.
Stracquadanio, M., Apollo, G. and Trombini, C. (2007). *Water, Air, and Soil Pollution*, 179, pp. 227-237.
Strandberg, B., Sunesson, A.L., Sundgren, M., Levin, J.O., Sallsten, G. and Barregard, L. (2006). *Atmospheric Environment*, 40, pp. 7686-7695.
Tang, N., Hattori, T., Taga, R., Igarashi, K., Yang, X.Y., Tamura, K., Kakimoto, H., Mishukov, V.F., Toriba, A., Kizu R. and Hayakawa, K. (2005). *Atmospheric Environment*, 39, pp. 5817-5826.
Tasic, M.D., Rajsic, S.F., Novakovic, V.T., Mijic, Z.R. and Tomasevic, M.N. (2005). *Physica Scripta,* T118, pp. 29-30.
Wojas, B. and Almquist, C. (2007). *Atmospheric Environment*, 41, pp. 9064-9078.

3.4 Modelling air quality

During recent decades mathematical modelling has become a widely used tool in assessing urban air quality. Mathematical models can be especially useful in predicting air pollution levels, for example in forecast systems for a large number of places, as required in population exposure studies; or as to inform political decisions in assessing future air quality scenarios. Some users demand instantly accessible online information on air quality. Such users include research institutes which need effective methods and tools, firstly for monitoring the exposure of individuals to pollutants, and secondly for the back tracking of pollution sources. Historical databases are used in these simulations. For assessing exposure in environmental-epidemiological studies, access to a range of powerful software is required. There is also demand from the general public for real-time air quality information. Another important use of the system is to compare real-time air quality information with other urban areas in Europe.

One important task for the emission databases is to generate data for dispersion calculations. The description of the emissions and their variation in time and space has to be detailed in order to carry out good dispersion calculations. Some emission sources are not included, but are instead described as background concentrations in large-scale modelling.

The several dispersion models and data processing methods of the system use emission data, and monitored meteorological data. Both vertical and horizontal distribution of air pollutants is calculated. The dispersion models allow studies of various hypothetical situations, such as increased traffic flow, changed climatic conditions and additional emission sources.

The science of modelling air quality and specific constituents is extremely sophisticated, particularly where adequate input data on meteorological, topographical and chemistry constituents is available. Many urban areas have detailed air modelling capacities, traditionally because of such serious air pollution problems. The classes of models include:

- Simple deterministic
- Statistical
- Box and multibox
- Local plume and puff
- Finite difference and grid
- Particle
- Physical wind tunnels Regional
  (As categorized by Szepesi, 1989)

Box/multibox or Gaussian models are the most common models for air quality estimation in urban areas.

3.4.1 Box models

This is a common simple model used to get an initial estimate of concentration values. It is based on the mass conservation of a pollutant in a box. The reference frame is Eulerian i.e. "fixed frame", rather than Lagrangrian, i.e. using a reference frame that moves with the velocity of the pollutant. The box or volume may represent a city or region. The plan area over a city is represented by $\Delta x \Delta y$. Δz is the vertical dimension of the airshed.

Fig. 9: Schematic of a box model in the atmosphere

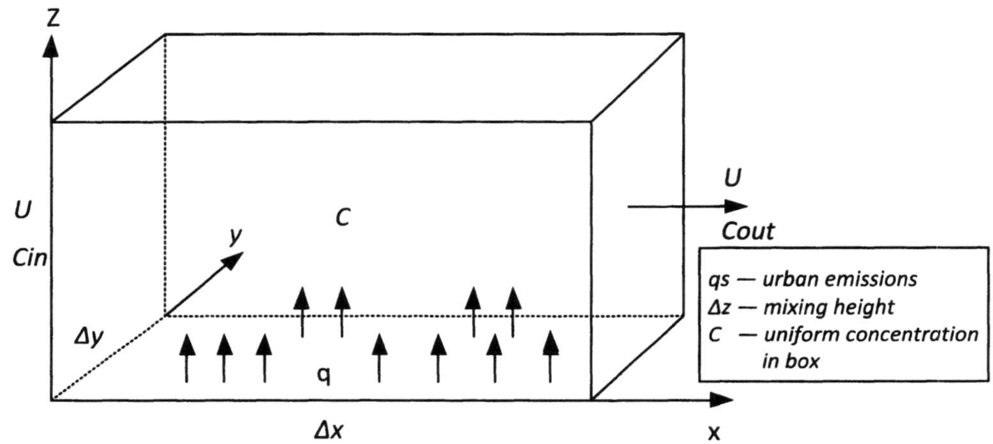

Consider the wind entering the airshed with a velocity U and a concentration C_{in}. Assume no pollutant leaves the side walls of the box, and that full mixing occurs within the box. For the sake of simplicity, the pollutants are assumed to be conservative (i.e. the generation/decay rate $\rightarrow 0$):

Rate of change of pollution in box = rate of pollution entering – rate of pollution leaving + rate of pollution generation or decay

$$\Delta x \Delta y \Delta z \frac{\partial C}{\partial t} = \Delta x \Delta y q_s + \Delta y \Delta z U (C_{in} - C_{out}) + \Delta x \Delta y \Delta z r_g \tag{5}$$

For no generation within the box, $r_g=0$,

$$\frac{\partial C}{\partial t} = \frac{q_s}{\Delta z} + \frac{U}{\Delta x}(C_{in} - C_{out}) \tag{6}$$

Assuming constant concentration throughout the box, then C_{out} = C, and taking the limit,

$$\frac{dC}{dt} = \frac{q_s}{H} + \frac{U}{L}(C_{in} - C) \quad (7)$$

where $\Delta x = L$ – length of box

$\Delta z = H$ – mixing height.

The steady state case reduces to

$$C = \frac{q_s L}{UH} + C_{in} \quad (8)$$

The unsteady case becomes

$$\frac{dC}{dt} + \left(\frac{U}{L}\right)C = \frac{q_s}{H} + \frac{U}{L}C_{in} \quad (9)$$

$$\frac{L}{U}\frac{dC}{dt} + C = \frac{q_s L}{UH} + C_{in} \quad (10)$$

And the analytical solution is:

$$C(t) = \left(\frac{q_s L}{UH} + C_{in}\right)\left(1 - e^{(-Ut)/L)}\right) + C(0)e^{(-Ut)/L} \quad (11)$$

If there is only pollution entering the box from city emissions and the initial concentration in the box is zero, the Eq. (11) is than simplified to:

$$C(t) = \frac{q_s}{UH}(1 - e^{(-Ut)/L}) \quad (12)$$

Multibox models are extensions of the box model. The air pollutants are assumed to be well mixed in each box, and each box is linked. Within a box, reaction and removal processes are allowed. Multibox models give better resolution in time and space than a single box. At box junctions, the interactions between boxes are defined as in boundary-type conditions.

3.4.2 Gaussian dispersion model

The Gaussian plume model for single and multiple sources is widely used for air pollutant models. Equation 13 describes the three-dimensional concentration field generally by point source(s) under stationary meteorological and emission conditions:

$$C(x,y,z)=\frac{Q}{2\pi\sigma_y\sigma_z u}\exp\left[-\frac{1}{2}\left(\frac{y}{\sigma_y}\right)^2\right]\left\{\exp\left[-\frac{1}{2}\right]\left(\frac{z+H}{\sigma_z}\right)^2+\exp\left[-\frac{1}{2}\left(\frac{z+H}{\sigma_z}\right)^2\right]\right\} \quad (13)$$

where *C(x,y,z)* – pollutant concentration,
H – height of the emission,
Q – emission rate,
$\sigma_y\sigma_z$ – standard deviation of the plume concentration distribution along *y* or *z*.

In many cases, including traffic, fugitive gases or smouldering landfill sites at ground level equation reduces to:

$$C(x,y,z)=\frac{Q}{2\pi\sigma_y\sigma_z u}\exp\left[-\frac{1}{2}\left(\frac{y}{\sigma_y}\right)^2\right]\exp\left[-\frac{1}{2}\left(\frac{H}{\sigma_z}\right)^2\right] \quad (14)$$

If we deal with the road traffic emissions, we will treat this as several point emissions. Depending on the emission characteristics, we estimate an initial distribution of the pollutant defined by the initial (at X=0) values of σ_y and σ_z. (Airviro, 1997). The values σ_y and σ_z are the Pasquill-Gifford parameters and depend on the atmospheric stability class and the distance from the source. The modelling of these equations, with facilities for multiple sources and complex topography, is a feature of many proprietary computer models. However, as the model is based on stationary meteorological and emission conditions, it is not usable in pollution/air quality forecasting. Dry deposition, wet deposition and chemical transformation processes can be accounted for in the Gaussian models (Zanetti, 1990) by multiplying basic equation (14) by exponential terms such as

$$\exp\left(-\frac{t}{T}\right)$$

where, $t=\frac{x}{u}$ – travel-time,
T – timescale.

Timescales for dry deposition can be expressed as

$$T_d=\frac{\Delta h}{v_d} \quad (15)$$

where Δh – vertical plume thickness,
v_d – deposition velocity.

Dry deposition does not occur within the immediate downwind distance of the emission, and Zanneti suggests that equation (14) be applied further downwind of x_d where:

$$x_d = \frac{H_e}{2\sigma_z},$$

where H_e – plume effective length.

Similarly, the timescale for wet deposition is expressed as

$$T_w = \frac{3.6 \times 10^6 P_L}{S_r P_R} \tag{16}$$

where P_L – thickness of precipitation layer,
S_r – scavenging ratio,
P_R – precipitation rate.

The timescale for chemical transformation depends on a number of factors: whether the pollutant is organic or inorganic; the temperature; the gas solubility, etc. General air pollution modelling tends to operate on the basis of conservative pollutants, but most pollutants are reactive. The time-scale for the transformation of SO_2 is around 24 hours.

Gaussian models can be spatially integrated to simulate the effects of line sources (traffic on a street or highway). In these cases integration would be numerical rather than analytical.

3.4.3 The Grid model

The Eulerian Advection-Diffusion grid dispersion model is used to meet the requirements given by areas with a complicated topography. The model has some additional advantages compared to the standard Gaussian dispersion approach:

- it allows transient simulations and can simulate the accumulation of pollutants during calm wind conditions;
- boundary conditions (such as deposition) as well as particle characteristics (such as settling) may be introduced.

The Grid model does not resolve individual buildings. Instead, surface structures like houses and trees enter the model through local roughness values and, as a consequence, through the wind field. Buildings give a rougher surface; more friction means a lower wind speed, which influences dispersion, and so on.

The default setting for the calculation height is 2 m, but the user may specify any other value. The height at which the simulated concentration values are valid will coincide with the calculation height above ground in open areas (i.e. in the countryside). Over a city area, the simulation will reflect the concentrations at the specified height above the roof height. The default setting will thus give concentrations at roof height, not the concentration at street level. The blocking effects of buildings surrounding a street, resulting in raised concentration levels, can be studied separately.

The Grid model is designed for use on larger areas. It is not suitable for use in calculations on a small scale.

The model is based on the three-dimensional advection-diffusion equation for a certain substance or non-interactive particles (Pielke, 1984):

$$\frac{\partial c}{\partial t}+u\frac{\partial c}{\partial x}+v\frac{\partial c}{\partial y}+w\frac{\partial c}{\partial z}=\frac{1}{\rho}\frac{\partial}{\partial z}\left[\rho\left(K_z\frac{\partial c}{\partial z}+w_s c\right)\right]+\frac{Q}{\rho}; \quad (17)$$

where c – concentration, mass of substance per unit volume/mass of air per unit volume
ρ – density of air, a function of height
u,v,w – wind velocity components,
x,y,z – Cartesian coordinates,
w_s – particle settling velocity,
Q – sources,
K_z – vertical turbulent diffusion coefficient.

The equation is solved on a three-dimensional numerical grid. Two successive coordinate transformations of the equation are made. The necessary three-dimensional wind field (u,v,w) is produced. Above the surface layer, an algebraic formula is used for all stability conditions.

$$K_z(z)=K_z(0.1mixh)+\frac{(z-mixh)^2}{(mixh-0.1mixh)^2}[K_z(0.1mixh)-K_z(mixh)+$$

$$(z-0.1mixh)\left(\frac{\partial K_z}{\partial z}(0.1mixh)+2\left(\frac{K_z(0.1mixh)-K_z(mixh)}{(mixh-0.1mixh)}\right)\right)] \quad (18)$$

In order to create a three-dimensional wind field – necessary for the three-dimensional grid model – vertical profiles of the horizontal wind field are calculated at every grid point. The profiles are calculated according to methods described in the "Wind fields" section. The profiles are distributed through the boundary layer up to the height *mixh*, the mixed layer height. Above *mixh*, a free wind field, constant up to the top of the calculation domain, is assumed to prevail.

3.4.4 Street Canyon model

Both the Gauss and the Grid model calculate roof level concentrations. The Street Canyon model is a small-scale model that allows you to simulate the street level concentrations on a single road that has a row of buildings on each side. The simulated concentrations are visualised on a vertical plane perpendicular to the street, the buildings creating the effect of looking down a canyon.

This model simulates the dispersion from traffic on a particular road. The user must be aware of the fact that traffic on other roads may very likely contribute to a substantial part of the pollution levels, especially if the simulated road has a low traffic density and is surrounded by larger streets. Will it then be possible to separately estimate the fractions of street pollution created by the local traffic and those created by emissions on other places?

The basis of the model comes from studies in San José, USA, where measurements indicated a vertical eddy circulation when the wind was blowing perpendicular to the street. The eddy circulation implies higher pollution concentration on the upwind side of the road, because this air includes the traffic exhausts. The leeward side shows much lower concentrations, as the air originates at roof level. The model is referred to as the “Standford model”.

The “Stanford model” expressions have been tested and modified in a study in Scandinavia. The expressions for the *CL* (concentration leeward side, i.e. on the sheltered side of the street) and *CW* (concentration on the windward side) are calculated:

$$CL = \frac{K \cdot Q}{(u+0.5) \cdot \left[\left(x^2 + y^2 \right)^{\frac{1}{2}} + L_0 \right]}, \quad (19)$$

$$CW = \frac{K \cdot Q \cdot (H - z)}{W \cdot (u + 0.5) \cdot H}, \quad (20)$$

where K – empirical constant,
Q – traffic emissions ,
u – wind component perpendicular to the street axis,
L_0 – length scale of the individual cars,
W – width between the houses,
H – typical house height,
x, z – horizontal and vertical distances from the street emission segments.

Fig. 10: Principles of air circulation within a street canyon (the upwind side is denoted UW, the leeward side LW, the road width RW).

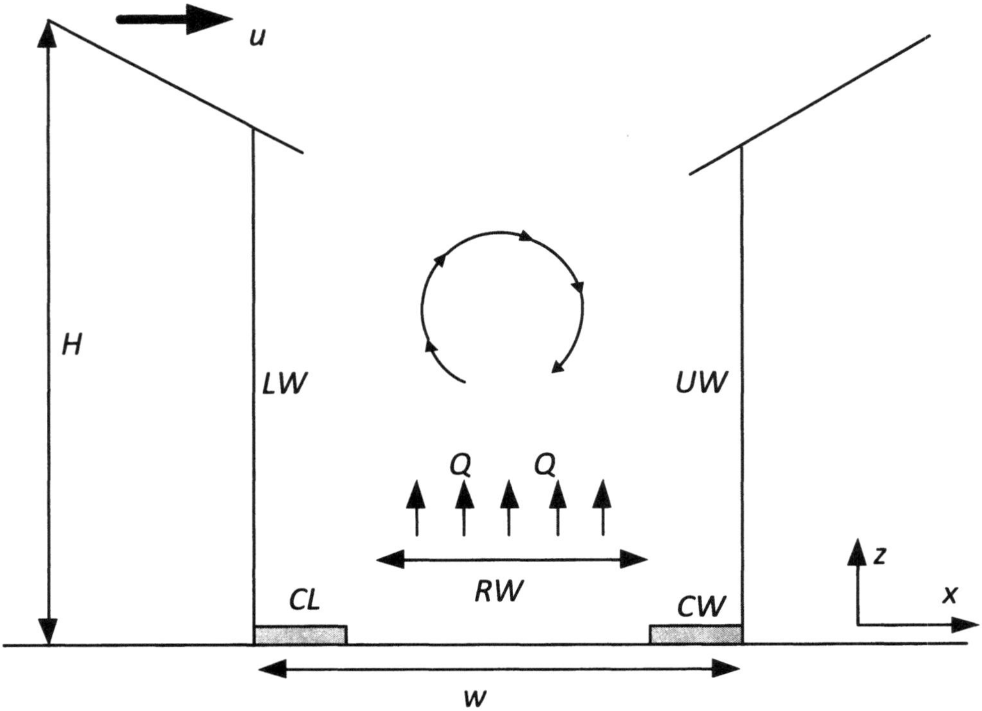

The vertical section between the houses is divided into a grid, on which the concentrations are evaluated. The road, i.e. that section of the street width – *RW* – where the emissions originate, is divided into a large number of road segments with equal emission rates.

The wind component is taken from the wind field calculation, evaluated at the street location. If the wind is blowing within 22.5 degrees from the angle of the street, the concentration will be found as an average of *CL* and *CW*, since they are symmetrical.

For cases where the wind direction is closer to being perpendicular to the street, we will find an asymmetry, with higher concentrations on the sheltered side of the road (leeward side) and lower values on the opposite side (windward side). The concentrations on the windward side are calculated according to *CW*. At the other gridpoints, a summation of *CL* values takes place, implying a contribution from each emitting segment situated upwind of the gridpoint itself (there is no impact from an emitting segment situated downwind).

3.4.5 Wind fields

The wind field calculation is based on the concept first described by Danard (1977), where mesoscale winds are generated by using:

- a horizontal momentum equation,
- a pressure tendency equation,
- a first thermodynamic equation.

This concept assumes that small-scale winds can be seen as a local adaptation of large-scale winds (free winds), due to local fluxes of heat and momentum from the sea or the Earth's surface. Any non-linear interaction between the scales is neglected. Danard assumes that the adaptation process is very fast: 1.5 hours for model resolutions of 10*10 km. It is also assumed that horizontal processes can be described by non-linear equations, while the vertical processes can be parameterised as linear functions. The large-scale winds, as well as vertical fluxes of momentum and temperature, are estimated from profile measurements in one or more meteorological masts (called principal masts). When the topography is relatively smooth – without dominating ridges or valleys – the free wind is assumed to be horizontally uniform. For deep bi-directional valleys, this is not possible. A channel flow approach has been used to include the effects governing the free winds in deep valleys. For larger areas, or for meteorologically complex areas (for example where sea breezes are encountered), information from more than one meteorological station is needed. The data from these stations will then pass through a mesoscale interpolation before entering the wind model.

Stability and turbulence conditions in the boundary layer are evaluated at the locations of the principal masts, i.e. where information about both the temperature gradient and the wind speed at one or two levels is available. This data is used to calculate the boundary layer scaling parameters.

The two most important parameters in terms of the scaling of the atmospheric stability and turbulence in the surface boundary layer are the *vertical heat flux H* and the *friction velocity u_**. From these, a characteristic length scale – the Monin-Obukhov length – can be calculated:

$$L = \frac{\overline{T}}{g} \frac{u_*^3 \rho c_p}{\kappa H}, \tag{21}$$

where $\overline{T}$ is the average surface air temperature, g is gravity, κ is the von Kárman constant and ρc_p is the specific heat capacity. The determination of the Monin-Obukhov length L follows the profile method discussed by Berkowicz and Prahm (1982 a,b).

Using the recorded potential temperature difference Δq (close to the ground) and wind speed U (at one or two levels), L is calculated using an iterative method. The local value of the ground roughness length z_0 must also be estimated (Panofsky and Dutton, 1984). If both the temperatures and one or two velocities are available, it is possible to calculate two other scaling variables: u_*, T_* – temperature scale:

$$u_* = \frac{\kappa\left(U_{z2} - U_{z1}\right)}{\ln\left(\frac{z_{u2}}{z_{u1}} - \Psi_{m2} + \Psi_{m1}\right)}, \tag{22}$$

$$T_* = \frac{H}{\rho c_p u_*} = \frac{\kappa \Delta\theta}{\ln\left(\frac{z_{t2}}{z_{t1}} - \Psi_{H2} + \Psi_{H1}\right)}, \tag{23}$$

The measurement heights are z_{u2}, z_{u1}, z_{t2} and z_{t1}. If wind is measured at only one level, z_{u2}, z_{u1} is set equal to z_0 and U_{z1} to 0.The similarity functions Ψ_{m2}, Ψ_{m1} are functions of z/l. For stable conditions ($L>0$) we can use (Dyer, 1974):

$$\Psi_m = -5 \cdot \frac{z}{L},$$

$$\Psi_H = -5 \cdot \frac{z}{L}.$$

For unstable conditions ($L<0$), Paulson (1970) proposes

$$\Psi_m = \ln\left[\left(\frac{1+x}{2}\right)^2 \left(\frac{1+x}{2}\right)^2\right] - 2\arctan(x) + \frac{\pi}{2}, \tag{24}$$

where

$$x = \left(\frac{1-16z}{L}\right)^{\frac{1}{4}},$$

and

$$\Psi_H = 2\ln\left[\frac{1}{2}\left(1+\sqrt{1-\frac{16z}{L}}\right)\right]$$

Fig. 11: The Monin-Obukhov length L behaviour as a function of stability

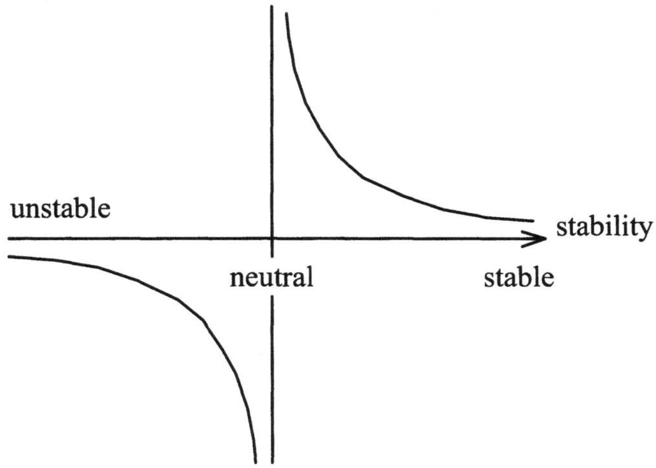

The calculated Monin-Obukhov length L is further used to classify the stability conditions. The depth of the boundary layer Zi is a key parameter, since the pollutants released close to the ground will be more or less trapped inside this layer.

For stable conditions L>0 the Zilitinkevich (1972) estimation is:

$$Zi = 0.4\sqrt{\frac{u_*}{f}L}\,, \tag{25}$$

This estimate relies on the assumption that the inversion is caused by local cooling. It will not serve for cases with relatively high inversions that are caused by advective air masses.

The corresponding expression for neutral and unstable conditions L<0 is (Panofsky and Dutton, 1984):

$$Zi = 0.3\left(\frac{u_*}{f}\right).$$

In practice, the mixing height *mixh* often cannot be set equal to the boundary layer height (Zi). Usually it is derived from the time series and varies with latitude. Direct measurements of velocity fluctuations may be used to evaluate the dispersion coefficients σ_y and σ_z, which regulate lateral and vertical diffusion of the Gaussian plume. The quantities recorded are the standard deviation of wind direction σ_a and vertical wind speed σ_w.

3.4.6 Aerosol dynamics modelling and timescales

The processes acting on atmospheric aerosols are listed in the following table and their impact on the particle number and volume is briefly explained.

Table 9: Processes that alter the particle size distribution in the atmosphere and their effect on the particle number dN and particle volume dV, "+" means increase, "–" means decrease, while "0" stands for no change (adapted from Ketzel, 2004)

Process	**Sink/source**	**dN**	**dV**
Nucleation	source of new, very small particles	+	((+))
Condensation	gas to particle conversion	0	+
Evaporation	caused particle to shrink and eventually disappear totally	0 or –	–
Coagulation	particles collide with each other and sink together	–	0
Dry deposition	sink of particles by surface contact	–	–
Emission	source	+	+
Dilution, mixing	source or sinking depending on the concentration of the diluting air mass	+ or –	+ or –

A detailed description and mathematical formulation of the different processes is presented in several works (e.g. Seinfeld and Padis, 1998; Hinds, 1999).

Similarly to dispersion modelling, a great variety of aerosol dynamics models exists, varying widely in complexity. The models differ for example in the size of modelling domain, the description of the aerosol size distribution, the mixing state (externally or internally mixed), chemical composition, and in the number and accuracy of the aerosol dynamical and chemical processes included. The range begins with, for example, a simple box model describing the development of the monodisperse aerosol over time, and extends to regional or global models with a representation of single particles, each with a different composition and size (Kleeman and Cass, 2001).

In measuring the *coagulation* caused by the Brownian motion of the particles, thermal coagulation, an ever-present process, needs to be taken into account. Kinetic coagulation occurs as a result of velocity differences between particles, which arise due to differing settling velocities, a strong flow gradient, or in very turbulent flow. Hinds (1999) shows that kinematic coagulation may attain the efficiency of thermal coagulation in extremely turbulent or shear flow; and that this is true only for particles larger than several μm in size. Wexler et al. (1994) conclude that coagulation in the atmosphere is clearly dominated by Brownian motions, except for instance turbulence that may occur in clouds.

Nucleation and condensation of sulphuric acid and semi-volatile organic substances (e.g. unburned fuel and lube oil) is especially responsible for the formation of the < 50nm in vehicle exhaust during the first milliseconds of dilution (Shi et al., 1999; Sakurai et al., 2003; Sipilä, 2010). Particle nucleation events and subsequent growth are frequently observed in clean and polluted natural air masses, and this is currently the subject of intensive research.

The *timescales* depend on the "lifetime" of the exhaust particles and on the particle size. Before including a description of the processes in an urban dispersion model, it is worthwhile to analyse the importance of the different processes on the street scale. An extremely useful way to compare different sources, sink and transformation processes acting on compounds in the atmosphere is to evaluate the timescale of each process. The process with the smallest timescale is most relevant in changing the concentrations of the compound. The often-used term "resistance time" is based on the same concept. The timescale τ can be expressed as a ratio of the concentration and the change in concentrations due to the process under examination:

$$\tau_{processX} = \left|\frac{C}{\dot{C}}\right| ; \ \dot{C} = \frac{\delta C}{\delta t}_{process\,X}$$

References

Airviro user documentation, SMHI (1997).

Berkowicz, R. and Prahm, L.P. (1982). Evaluation of the profile method for estimation of surface fluxes of momentum and heat. *Atmospheric Environment*, 16, pp. 2809-2819.

– (1982). Sensible Heat Flux Estimated from Routine Meteorological Data by the Resistance Method, J. *Applied Meteorolo*gy 21, pp. 1845-1864.

Danard, M.B. (1977). A simple model for mesoscale effects of topography on surface winds. *Monthly Weather Review,* 105, pp. 572-581.

Dyer, A.J. (1974). A review of flux-profile relationships. *Boundary Layer Meteo.* 7, pp. 363-372.

Hinds, W.C. (1999). *Aerosol Technology*. John Wiley and Sons, New York.

Ketzel, M. (2004). *Dispersion and Transportation of Traffic Exhaust Particles in the Urban Atmosphere.* PhD Thesis, Lund University.

Kleeman, M.J. and Cass, G.R. (2001). A 3D Eulerian source-oriented model for an externally mixed aerosol. *Environmental Science & Technology* 35, pp. 4834-4848.

Panofsky, H.A. and Dutton, J.A. (1984). *Atmospheric Turbulence.* Wiley, p. 397.

Paulson, C.A. (1970). The mathematical representation of wind speed and temperature profiles in the unstable atmospheric surface layer. *J. Appl. Meteor.,* 9, pp. 857-861.

Pielke, R.A. (1984). *Mesoscale meteorological modelling.* Academic Press, ISBN 0125548206.

Seinfeld, J.H. and Pandis, S.N. (1998). *Atmospheric Chemistry and Physics: From Air Pollution to Climate Change.* John Wiley and Sons, New York.

Shi, J.P., Harrison, R.M. and Brear, F. (1999). Particle size distribution from a modern heavy duty diesel engine. *The Science of The Total Environment* 235, pp. 305-317.

Sipilä, M., Berndt, T., Petäjä, T., Brus, D., Vanhanen, J., Stratmann, F., Patokoski, J., Mauldin, R.L. 3rd, Hyvärinen, A.P., Lihavainen, H. and Kulmala, M. (2010). The role of sulfuric acid in atmospheric nucleation. *Science,* 327(5970), pp. 1243-6.

Szepesi, D.J. (1989). *Compendium of Regulatory Air Simulation Models*, Akademiai Kiado, Budapest.

Wexler, A.S., Lurmann, F.W. and Seinfeld, J.H. (1994). Modelling urban and regional aerosols: I. Model development. *Atmospheric Environment* 28, pp. 531-546.

Zannetti, P. (1990). *Air pollution modelling: theories, computational methods and available software.* New York, Van Nostrand Reinhold.

Zilitinkevich, S.S. (1972). *On the determination of the height of the Ekman boundary layer.* Bound.-LayerMeteor., 3, pp. 141-145.

3.5 Case study II. Ambient air quality mapping for integrated urban traffic flows management

***Source**: Kliučininkas L., Matuzevičius D., Mockuvienė J. (2006). Ambient Air Quality Mapping for Integrated Urban Traffic Flows Management. Integrative Approaches Towards Sustainability in the Baltic Sea Region. Peter Lang GmbH, ISSN 1434-3819, vol. 21, pp. 465-471.*

This study addresses issues related to the assessment of the environmental quality in order to support traffic flows management. This investigation gives an example of the formation of seasonal air quality maps. The formation of maps integrates air quality monitoring data, intensity counts of traffic flows, and estimations of meteorological conditions. The AIRVIRO v. 2.20 software is applied to model chosen weather conditions. To address traffic flows management and minimise air pollution in Kaunas , ambient air quality modelling/mapping based on conceptual scenarios has been applied. User-defined scenarios have been applied in pursuit of an appropriate transportation infrastructure.

Introduction

Rapidly growing urban traffic flows have been identified as major contributors to the ambient air quality in a number of medium and large cities. Over the last decade, many cities in the Central Eastern European (CEE) countries have had to face the consequences of a substantial increase in private car ownership and a public transport crisis. Most of these cases involved used cars from Western Europe, imported into the CEE countries.

Central Kaunas represents a hub for administrative, commercial and cultural activities, and is heavily affected by traffic flows. Traffic Jams have become an endemic feature of the city centre due to insufficient street capacity. The majority of the 3-5-storey buildings in the city centre create unfavourable conditions for the distribution of pollutants.

The integration of data on traffic flows and air quality through modelling/ mapping systems is considered crucial for efficiency in transport systems planning. The major objective of such systems is to provide analytical results for urban planners and decision-makers, in order to meet current requirements of city development. Mapping and scenario results can be compared with air quality limits. An impact assessment of traffic air pollution abatement measures can also be carried out (Jensen et al., 2001). The probabilistic mapping technique was applied to review of air quality in central London in 1996/97 and predict main air quality parameters as set by air quality objectives in 2005 (Colvile et al., 2002). The British authors report on the development of a programme designed to provide detailed estimates of vehicle emissions on urban road networks, and so act as a precursor to urban air quality modelling (Namdeo et al., 2002). The programme includes emission and air quality mapping, an evaluation of associated transport policies and scenarios, and the preparation of inputs to other (e.g., epidemiological) models.

Measurements and basic data properties

Six sampling sites (see Fig. 1) were selected to carry out measurements relevant to modelling the impacts of traffic flows on ambient air quality. The sites were situated in a radius of 2.5 km to examine the spatial pollution variations within the neighbourhoods. The results from five sampling sites (No. 1-3,5,6) were obtained using a mobile laboratory; sampling site No. 4 represented a permanent monitoring station. To conform with EU siting criteria, sampler inlets were placed at least 2.3 m above ground level. The sampling sites were primarily selected to obtain representative measurements from the study area; other considerations included the security of equipment and the availability of an uninterrupted power supply.

Fig. 1: The locations of the sampling sites

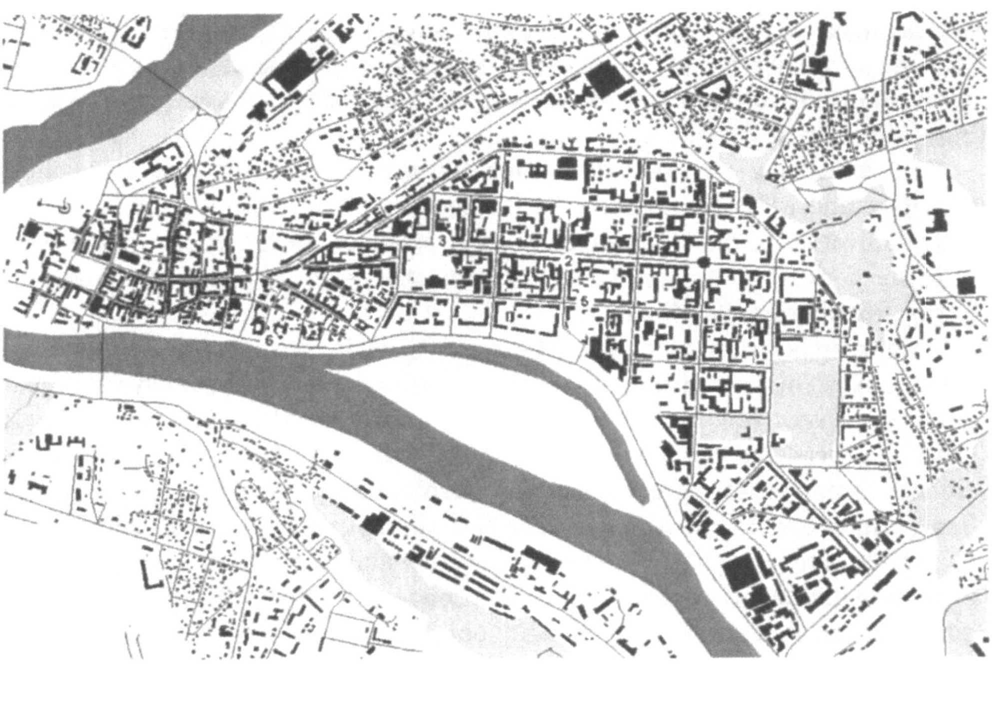

The measurement installations from Environnement S.A. (Paris-Poisy, France) comprised a chemiluminescent NO/NO_X/ NO_2 analyser (model AC31M) andan IR gas filter correlation CO analyser (model CO11M). The error-margin for the AC31M analyser is 0.17 ppb; for the CO11M it is 0.05 ppm. *In situ* measurements for selected sites were conducted during working days between 9 a.m. and 5 p.m. Resultant *in situ* data was obtained from the hourly averages of the continuous measurements. The 24-hour continuous sampling data from the stationary monitoring station was averaged on a half-hour basis. Table 1 presents descriptive features of measurement sites with respect to the degree of impact.

Table 1: Measurement sites

No	Site	Type of impact
1	University, main building	35 metres away from K. Donelaicio street. Moderate impact from the street
2	University, Faculty of Management	160 metres from Kestucio street. No. direct impact from the street
3	Municipality	165 metres from K. Donelaicio street. No. direct impact from the street
4a	Monitoring station *Centras*	12 metres from the street. Strong impact from the street
4b	Bank	4 metres from St. Gertruda street. Very strong impact from the street
5	Police station	4 metres from Kestucio street. Very strong impact from the street
6	Fire brigade	4 metres from King Mindaugas street. Very strong impact from the street

The urban background concentrations of NO_X and CO were estimated using data series from the *Silainiai* stationary monitoring station. The station was situated in a highly populated district of Kaunas and was not directly influenced by emissions from transport and industries. The urban background concentration of NO_X in the city was 15 $\mu g/m^3$. The urban background concentration of CO was estimated at 0.05 mg/m^3 level.

Formation of seasonal ambient air quality maps

Model

The Airviro software developed by the Swedish Meteorological and Hydrological Institute (Airviro 2.20) integrates different functions required in Air Quality Management. In the central part of Kaunas, the dominant part of the air pollution is induced by the street traffic emissions. Emissions from the streets were assumed to be evenly distributed point emissions along the lines and were a good approximation of the real situation (Russel et al., 1983). In order to understand how air quality in the central part of Kaunas will be affected by changes in meteorological conditions, authors have applied the following Airviro functional blocks – Emission Database, Dispersion Modelling and Indico data analysis and presentation package.

The Traffic Emissions Database comprises data on traffic speed, traffic intensity and variations as well as information on the types and proportions of vehicles operating on that street. Consequently, the variations as well as proportions of different vehicles on a street could be defined by the structural scheme presented in Fig. 2.

Fig. 2: Structural scheme of traffic emission database

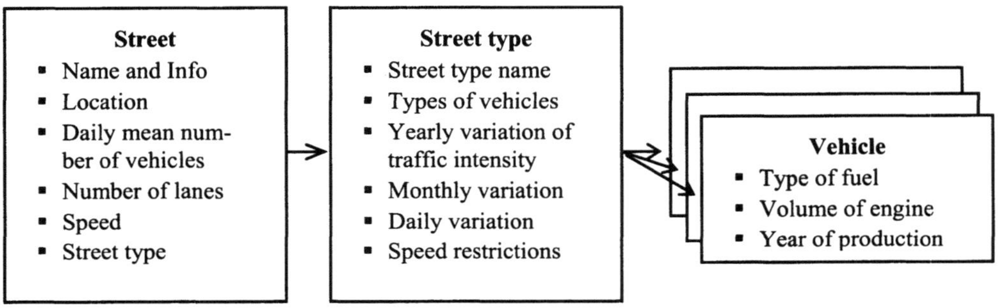

For dispersion modelling, authors have applied the Gauss model, based on a Lagrangean-Gaussian formulation. For linear emission sources (streets) introduced individually, the horizontal dispersion coefficient σ_y is set to

$$\frac{dxy}{2},$$

where

$$dxy = \sqrt{width^2 + length^2}\ , \tag{1}$$

However, in built up areas where the Gauss model calculations are evaluated at roof level, σ_y is set to 0.667* building height.

The vertical dispersion coefficient σ_z is defined with an empirical formula and in open areas is set to 3.57-0.53 U, where U – horizontal wind velocity (Airviro, 1998).

Meteorological conditions

When forming the seasonal pollution dispersion maps, the meteorological conditions were considered driving variables of the model. The complex meteorological observations were performed in the vicinity of Kaunas.

The wind field calculations are based on the concept first described by Danard, where small-scale winds can be seen as a local adaptation of large-scale winds (free winds) due to local fluxes of heat and momentum from the sea or Earth's surface. The two most important parameters for the scaling of atmospheric stability and turbulence in the surface boundary layer are the vertical flux H and the friction velocity u. From these, a characteristic length scale – the Monin-Obukhov length is calculated:

$$L = \frac{\bar{T}}{g} * \frac{u^3 \rho c_p}{\kappa H}; \tag{2}$$

where $\bar{T}$ is the average surface air temperature, g is gravity, κ is the von Karman constant and ρc_p is the specific heat capacity. In the Airviro dispersion model, the Monin-Obukhov length is extensively used as a discriminator of different meteorological regimes (Panofsky et al., 1984).

The boundary layer stability is estimated in the form of calculated 50 and 98 percentiles of Monin-Obukhov length. The calculated Monin-Obukhov length (L) was used to classify the stability conditions in the boundary layer (see Fig. 3). Corresponding computations for chosen weather and emission scenarios were carried out.

Fig. 3: Essential meteorological parameters used to develop seasonal air quality maps

Season	**Percentile**	**Monin-Obukhov length, m**	**Wind direction, degree**	**Mixing height, m**
Cold	50	2222	240	1666
	98	7	217	24
Warm	50	1786	264	840
	98	7	260	24

Emission factors for vehicles

To calculate emissions produced by the exhaust systems of road vehicles, authors have applied EMEP/CORINAIR emission factors. Emission factors were worked out by the members of the CORINAIR Working Group (Eggleston et al., 1993), taking into account the results of comprehensive studies carried out in France, Germany, Greece, Italy, the Netherlands and the United Kingdom. In addition, some data measured in Austria, Sweden and Switzerland were incorporated.

Street traffic flows

The street traffic flows analysis was performed by the Transport and Road Research Institute. The study has involved measurements of traffic intensity, composition of flows and vehicle speed (TKTI, 2001).

Analysis of the data sets revealed that the average traffic intensity (ATI) for the central part of Kaunas, based on a 24-hour time series, with the exceptional case of the summer vacation period, is more or less stable during the whole year. Analysis of street traffic flows with respect to driving speed has shown that characteristic speed ranges between 40-60 km/h. The stationary counter-classifier Marksman 660 (Golden River Traffic, United Kingdom) was used to identify 24-hour yearly flow averages in the different vehicle classification groups.

Mapping ambient air quality scenarios

Scenarios were applied to the central part of Kaunas and estimated predictive scenario with the new M.K. Ciurlionis bridge: reference scenario (undeveloped traffic infrastructure around the bridge); scenario with re-distributed traffic flows (developed traffic infrastructure around the bridge). The predictive mapping was based on perspective data of traffic flows made available by municipal enterprise "Kauno planas".

The initial NOx and CO concentration mapping (scenario 0) was based on recent traffic flows data. The regressive data analysis was used to verify the model. The linear regression equation $y = 0.9589x + 4.2318$ for NO_x, and, where applicable, equation $y = 0.118x + 0.2034$ for CO, were used to correct the model output data. Figures 3 and 4 illustrate the distribution of NO_x concentrations.

The comparative studies of NO_x maps for the warm period (April 1 to September 30) revealed that a major part of investigated area falls into the concentration interval 20-60 $\mu g/m^3$. The NO_x concentrations in the close vicinity of the main streets amount to 100-120 $\mu g/m^3$. Due to the considerable distances from the main streets, the areas with high concentration of pedestrians are not affected by the increased concentrations of NO_x.

The trends of NO_x concentration levels for re-distributed traffic flows have shown that in the vicinity of the main streets the concentrations would decrease to 80-100 $\mu g/m^3$;in the area around M.K. Ciurlionis bridge, concentrations would rise from 70 to 100 µg/m^3. At the same time, high concentration would cover a bigger area. This will also have an effect on the NO_x µg/m^3 concentration increase in the streets directly connected to the bridge.

Fig. 3: Distribution of NO_x concentration: present situation

As the verification of the model showed, the hourly street concentrations could considerably differ from those seasonal concentrations modelled by the authors, especially on the street crossings. Therefore, the above considerations should be taken into account when interpreting the ambient air pollution maps.

Fig. 4: Distribution of NO_x concentration: scenario with predicted traffic flows

Discussion and conclusions

Scenario comparison, with the emphasis on the changes in emissions and exposure, are often of far more value in a planning context than absolute values. The magnitude and location of change can convey information to planners and decision-makers, highlighting the consequences of trends and of planned actions, and focussing attention both on where interventions might have the greatest effect, and on where flexibilities may exist (Affum et al., 2003). Scenario management provides the user with capabilities to create a new scenario or to retrieve one which has been previously stored , and to define a set of policy measures to be examined. A "reference" or "zero-state" scenario is always present, serving as the basis for the creation of new scenarios. The definition of a new policy measure is carried out through interactive procedures based on series of maps. User interaction falls into three different functional groups: scenario management, scenario analysis and scenario comparison (Kliucininkas et al., 2006).

The authors have come to the following conclusions:

- The model proved to be a valuable instrument for spatial pollution distribution in ambient air quality evaluation;
- Airviro V.2.20 software is sufficiently precise in evaluation of distributed NO_x and CO concentrations; meanwhile the verification sites should be at least 25 m distance from the street;
- Since Lithuania has no emission factors relevant to the existing car fleet, further investigations into adapting EMEP/CORINAIR databases would be required;
- Mapping is a quick, accurate and perspective tool for urban traffic flows management.

References

Affum, J.K., Brown, A.L. and Chan, Y.C. (2003). Integrating air pollution modelling with scenario testing in road transport planning: the TRAEMS approach. *The Science of the Total Environment*, vol. 312, iss. 1-3, pp. 1-14.

Airviro User Documentation. *Swedish* (1998). Hydrological Meteorological Institute, Norrkoping, Sweden.

Colvile, R.N., Woodfield, N.K., Carruthers, D.J., Fisher B.E.A., Rickard A., Neville S. and Hughes A. (2002). Uncertainty in dispersion modelling and urban air quality mapping. *Environmental Science & Policy,* 5, pp. 207-220.

Eggleston, S., Gaudioso, D., Girisen, N., Jourmard, R., Rijkeboer, R.C., Samaras, Z. and Zierock K.-H. (1993). *CORINAIR Working Group on Emissions Factors for Calculating 1990 Emissions from Road Traffic*. vol. 1: Methodology and Emission Factors. Final Report, Document of the European Commission.

Jensen, S.S., Berkowicz, R., Hansen, H.S. and Hertel, O. (2001). A Danish decision support GIS tool for management of urban air quality and human exposures. *Transport Research Part* D. 6, pp. 229-241.

Kliučininkas, L., Matuzevičius, D. and Mockuvienė, J. (2006). Ambient Air Quality Mapping for Integrated Urban Traffic Flows Management. *Integrative Approaches Towards Sustainability in the Baltic Sea Region.* Peter Lang GmbH, ISSN 1434-3819, vol. 21, pp. 465-471.

Namdeo, A., Mitchell, G. and Dixon, R. (2002). TEMMS: an integrated package for modelling and mapping urban traffic emissions and air quality. *Environmental Modelling & Software,* 17, pp. 179-190.

Panofsky, H.A. and Dutton, J.A. (1984). *Atmospheric Turbulence*. John Wiley & Sons.

Russel, A., McRae, G.J. and Cass, G.C. (1983). Mathematical modelling of the formation and transport of ammonium nitrate aerosol. *Atmospheric Environment,* 17, pp. 949-964.

Transporto priemoniu eismo intensyvumo skaiciavimas automatiniais eismo intensyvumo skaitikliais-klasifikatoriais (Trafic intensity measurements using automatic counters-classifiers). Report (2001). Transport and Road Research Institute, Lithuania.

3.6 Modelling noise dispersion

Environmental noise is the noise experienced by people outside households. Noise experienced by people in the workplace or occupational environment is not considered in this chapter. Transport sources account for a substantial proportion of human health and well-being (WHO, 2009). Data delivered in accordance with the Directive on Environmental Noise (Directive 2002/49/EC) is available through the Noise Observation and Information Services for Europe (NOISE, 2011).

For describing the impact of noise on humans, the so-called Equivalent Sound Pressure Level (Leq) needs to be calculated; that is, the mean value of sound intensity over time expressed in decibels. However, Leq is not enough for the characterisation of environmental noise. It is equally important to measure and display the maximum values of the noise fluctuations (Lmax), preferably combined with a measure of the number of noise events (Ln). For most people, noise pulses are more annoying than a steady pulse of noise.

Approximately 40 per cent of the population living in the largest cities in the EU-27 may be exposed to long-term average road traffic noise levels exceeding 55 decibels (dB), and at night almost 34 million people may be exposed to long-term average road noise levels exceeding 50 dB. The WHO night noise guidelines for Europe recommend that people should not be exposed to night noise greater than 40 dB. Night-time noise levels of 55 dB, described as "increasingly dangerous to public health", should be considered as an interim target in situations where the achievement of the guidelines is not feasible (Eurostat, 2010).

According to a German Environmental Survey for Children, compared with children of a higher socio-economic status, children from families of low socio-economic status are more heavily exposed to traffic and annoyed by road traffic noise during the day (UAB, 2009). Urban air quality and noise often share a common source and may cluster spatially. There are examples of successful integrated approaches to reducing both local air pollution and noise levels.

The draft European Directive on the assessment and management of environmental noise aims to define a common approach for combatting the effects of exposure to environmental noise (Directive 2002/49/EC). It lays down a framework for determining exposure to environmental noise, making information on environmental noise and its effects available to the public, and adopting action plans. The action plans should address priorities that may be identified by the exceedance of any relevant limit value, or by other criteria chosen by the member states. The actions may, for example, include traffic planning, land-use planning, technical measures at noise sources, selection of quieter sources, reduction of sound transmission, and regulatory or economic measures or incentives. The objective is to prevent and reduce environmental noise where necessary, and particularly where exposure levels can include harmful effects on human health, and to preserve environmental noise quality where it is good.

In order to assess the number of citizens exposed to harmful traffic noise, outdoor measurements and dispersion modelling is required. Road traffic is responsible for the highest percentage of the European population exposed to noise levels greater than 55 dB(A) (Leq, 24-hour). In particular, the high proportions of Leq between 55 and 65 dB(A) can be attributed to the growth of traffic volume during recent years. This more than offsets any possible results of tightening noise controls as experienced for example in EC legislation. In addition, tyre/road noise is now the dominant source of road-traffic noise, particular in free-flowing traffic at speeds above 40 to 50 km/hour. Roughly speaking, each 25 per cent increase in vehicle numbers will result in a noise level increase of 1 dB, whereby one heavy goods vehicle is equivalent to about eight passenger cars.

3.6.1 Definitions

dB scale. Sound levels are expressed on a logarithmic scale in decibels (dB). The sound pressure level L_p is an indication of the strength of an acoustic pressure wave and calculated as

$$L_p = 10 \cdot \lg\left(\frac{p^2}{p_{ref}^2}\right), \qquad (26)$$

where p is the amplitude of the pressure wave and p_{ref} is usually equal to $2 \cdot 10^{-5}$ Pa. The acoustic energy of a sound wave is proportional to p^2. Such a logarithmic scale is chosen since it lies close to the human perception of the amplitude of acoustic waves.

Some rules of thumb concerning the perception of differences in L_p are as follows:

A difference of1 dB is hardly perceived by the human ear, while a difference of 3 dB, equal to a doubling of the acoustic energy, is observed as a small increase. A difference of 10 dB is perceived as a doubling of the loudness.

A-weigthing. The human ear does not have the same sensitivity for all frequencies in a sound. Sounds with a frequency of a few thousand Hertz are perceived louder than low frequency and very high frequency sounds. To take this difference in sensitivity into account, the sound registered by a microphone is filtered using an A-weighting filter. The resulting sound pressure level is expressed in dB(A).

Equivalent sound levels. Traffic noise levels usually vary quickly over time. The *equivalent* sound level, L_{eq}, is defined as

$$L_{eq} = 10 \cdot \lg\left(\frac{1}{T}\int_{t_1}^{t_2}\frac{\rho_{rms}^2}{\rho_{ref}^2}dt\right). \qquad (27)$$

This L_{eq} is defined as the sound level of a constant sound which carries the same acoustic energy as in the time-varying signal over the time period $T = t_2 - t_1$. The exact value of L_{eq} depends on the length of the time period T. Within the field of traffic noise calculations, another quantity commonly used is the "sound exposure level", or *SEL*, which is the total intensity (pressure squared) integrated over time. For a vehicle passing by, *SEL* is equal to Leq+10log(T).

LDEN. The concept of the equivalent sound pressure level is of particular importance, since it forms the base of the noise indicator L_{den} proposed by the EU noise directive. L_{den} is defined as:

$$L_{den} = 10 \cdot \lg\left[\frac{12}{24}10^{\frac{L_{day}}{10}} + \frac{4}{24}10^{\frac{L_{evening}+5}{10}} + \frac{8}{24}10^{\frac{L_{night}+10}{10}}\right], \qquad (28)$$

where L_{day}, $L_{evening}$ and L_{night} are the A-weighted long-term average sound levels (as defined in ISO 1996-2: 1987) determined over the day periods (12 hours, usually from 07:00 till 19:00), evening periods (4 hours, usually from 19:00 till 23.00) and night periods (8 hours, usually from 23:00 till 7:00), respectively. European Union member states may shorten the evening period by one or two hours and lengthen day/night accordingly. As seen from the formula above, the sound pressure level during the evening period is increased ("punished") by dB, and the level at night by 10 dB. In this way, the increase in the impact of noise during the night and the evening ("rest"-period) is accounted for to a certain degree.

The choice of L_{den} as a common indicator has some important consequences for what is expected from traffic flow models generating the basic data. Traffic data is needed for a 24-hour period (thus also during the night). Since there is some freedom in the choice of evening and night periods, traffic flow models should accommodate this by generating hourly traffic data, for example.

3.6.2 Single vehicle and traffic flow models

The road traffic noise originates from the three main noise sources:

- rolling noise due to the tyre/road interaction,
- propulsion noise caused by the driveline of the vehicle,
- aerodynamical noise due to the turbulent flow along the car body.

The total noise production is the energetic sum of these three contributions:

$$L_{W0}=\lg(10^{LWR/10}+10^{LWP/10}+10^{LWA/10}). \quad (29)$$

For aggregation to traffic flow over a line segment, at least the traffic intensity Q and average speed v for each vehicle category is needed. Accuracy can be improved if the distribution (in time) of vehicle speed per category is available. For urban driving conditions, a more detailed noise calculation requires also the distribution (in time) of acceleration/deceleration per vehicle category. Maximum detail is achieved when not only the distributions are available but also all information (namely vehicle type, speed, and acceleration/deceleration) for each vehicle separately.

The equivalent line-source power level per unit length for a road segment under study $L_{W,line.eq}$ is calculated as follows:

$$L_{W,line.eq.}=L_{W0}+10\cdot\lg\left(\frac{Q}{v}\right), \quad (30)$$

where L_{W0} is the sound power level obtained from the single vehicle noise model averaged over the time the vehicle is in the segment of the road considered. Depending on the available information, the contributions from either groups of vehicles with certain properties ($Q > 1$), or the contributions from each single vehicle ($Q = 1$) are to be summed energetically.

Depending on the type of traffic situation, more or less detail is necessary. For a motorway, characterized by high vehicle speed and little acceleration/deceleration, a minimum data set is usually sufficient. Conversely, modelling a road intersection needs acceleration/deceleration data and if possible individual vehicle data. Information of traffic composition and the distribution of speed and acceleration/deceleration data are important for accurate calculation.

In order to assess the number of citizens exposed to harmful traffic noise outdoor measurements and dispersion modelling is required. Road traffic is responsible for the highest percentage of the European population exposed to noise levels greater than 55 dB(A) (Leq, 24-hour). In particular, the high proportions of Leq between 55 and 65 dB(A) can be attributed to the growth of traffic volume during recent years. This more than offsets any possible results of tightening noise controls as experienced for example in EC legislation. In addition, tyre/road noise is now the dominant source of road-traffic noise, particular in free-flowing traffic at speeds above 40 to 50 km/hour. Roughly speaking, each 25 per cent increase in vehicle numbers will result in a noise level increase of 1 dB, whereby one heavy goods vehicle is equivalent to about eight passenger cars.

Box 2

Traffic noise survey in Kaunas

A traffic noise survey presented in this case study was carried out in one of the biggest city street crossings. Street crossing Savanoriu Ave and K. Petrausko-Utenos streets are in the densely populated residential area.

The **C**omputer **A**ided **N**oise **A**batement programme CADNA-A was used to evaluate noise dispersion (DataKustic, 1998). The calculations were based on the following parameters: traffic flows, landscape, buildings (height, number of floors), and speed limits.

Differences of traffic flows during day and night-time were calculated using equations 1-3:

$$DTV=b*Q_{15\text{-}19}; \quad (29)$$

$$D=0.06*DTV; \quad (30)$$

$$N=0.011*DTV; \quad (31)$$

where: DTV – number of cars per day; b – correction coefficient; $Q_{15\text{-}19}$ – number of cars (15.00-19.00); D – number of cars during the daytime (cars/hour); N – number of cars during the night-time (cars/hour).

Nine measurement points were chosen to verify the modelling results. Points M1, M2, M3, M4, M5 were situated in close vicinity to the streets; points M6, M7, M8, M9 inside housing blocks (see Fig. 12).

Fig. 12: Distribution of dwellings and measurement points

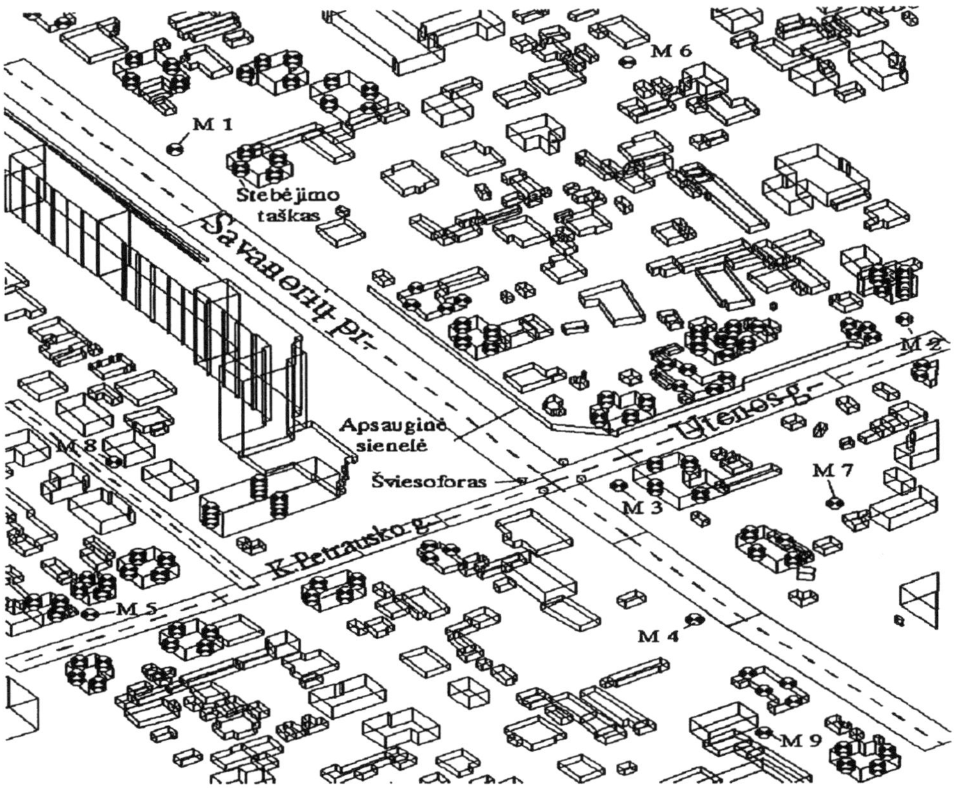

Fig. 13: Results of daytime noise dispersion modelling

The maximal noise level presence was modelled at the point closest to the road crossing. This level reaches 85 dBA, which is 20 dBA higher than maximum allowable noise levels.

Modelling effectiveness of noise abatement measures

After an analysis based on current calculated noise levels, different situations were simulated by changing traffic speed and flows. The initial speed limit was chosen as 60 km/h with asphalt concrete as the road surface. Later the speed limit was reduced to 50 km/h, then 40 km/h, and finally to 30 km/h. Afterwards, the following scenarios were estimated:

- Elimination of heavy transport;
- Elimination of heavy transport and microbuses;
- Elimination of heavy and public transport, as well as 25 per cent of individual cars;
- Elimination of heavy and public transport, as well as 50 per cent of individual cars;
- Construction of noise barriers.

Discussion

The modelling results (see Figure 13) show that the reduction of speed limits from 60 km/h to 50 km/h would help to reduce present noise levels by approximately 1.17 dBA; tol 40 km/h – by 2.36 dBA; and to 30 km/h – by 3.61 dBA. In the places with intensive traffic flows, speed-limit reduction is not an efficient scenario to achieve an allowable level of noise.

A more efficient noise reduction measure in the city could eliminate different categories of motor vehicles. The elimination of heavy transport (except public transport) would reduce present traffic noise levels by only 0.47 dBA. The elimination of heavy transport and microbuses is also not efficient. In this case, noise levels would be reduced only by 0.54 dBA. If some heavy and public transport, as well as 25 per cent of individual cars were eliminated, the present noise levels would be reduced by 4.2 dBA. Present noise levels could be reduced by 6.2 dBA, if heavy and public transport, as well as 50 per cent of individual cars, were eliminated.

Construction of noise barriers is quite expensive, but it is one of the most efficient noise reduction measures. The implementation of such barriers would be the most effective compared to other noise reduction scenarios.

***Source**: Taločkaitė E. , Kliučininkas L,. Zazerskis R. (2000). Traffic noise survey in Kaunas. Proceedings of the International Conference 'City Environment 2000', September 28-30, 2000, Kaunas, pp. 77-81.*

References

DataKustic (1998). *CADNA-A for Windows*, dataKustik GmbH, Munich, Germany, 422 p.

Directive 2002/49/EC of the European Parliament and of the Council relating to the assessment and management of environmental noise (2002) *Official Journal of the European Communities* L189, pp. 12-25.

Eurostat. (2010). Eurostat's population projection scenario – *EUROPOP2008*, convergence scenario.

Noise Observation and Information Service for Europe (NOISE) – http://noise.eionet.europa.eu/.

UAB (2009). The German Environmental Survey (GerES) for Children 2003/2006: *Noise. Environment & health* 01/2009, Dessau-Roßlau.

WHO (2009). Night Noise Guidelines. WHO Regional Office for Europe, Copenhagen.

3.7 Case study III. Noise mapping for the management of urban traffic flows

***Source:** Kliučininkas, L., Šaliūnas, D. (2006). Noise mapping for the management of urban traffic flows. Mechanika, vol. 3(59), pp. 61-66.*

This study analyses the EU policies related to assessment and management of traffic noise in urban areas. The analysis is based on the application of noise maps using *Cadna A* software for the central part of Kaunas in Lithuania. Data on the traffic flows, height of buildings, speed limits combined with actual noise measurements are used as an input to the dispersion model. The assessment of noise is based on the guidelines of the method 'XPS 31-133' defined in the European Commission Recommendation C(2003)2807. The analysis of noise maps, which is based on noise level measurement-modelling results, has raised the need for the quantitative determination the traffic flow type and the elaboration of acoustical equivalent for different modes of transport.

Introduction

Assessment and Management of Environmental Noise (European Commission, 2000) have been important issues in the European Parliament and the Council for the last decade. They have aimed to create the basis for a coherent and integrated EU policy. The Council has launched the idea of making EU-wide "noise maps" based on common methods and indicators. Noise mapping covers the whole mapping process, starting with the collection of raw data and the storage and retrieval of this data for computation/modelling, right up to the presentation of information related to outdoor sound levels, sound exposure and noise effects, or numbers of affected people. These maps should be available for the general public. They should form the basis for the development of action plans and strategies at local, national, and EU levels in order to combat noise pollution (Coelho, 2003).

The new EU Directive related to the assessment and management of environmental noise (Directive, 2002) requires a number of actions from the EU Member States to reduce noise levels. Noise maps and action plans are necessary for agglomerations with more than 250,000 inhabitants, major roads, major railways, and major airports. Articles 3, 5 and Annex I of Directive 2002/49/EC define the noise indicators L_{day} (daytime indicator), $L_{evening}$ (evening-time indicator), L_{night} (nighttime indicator), and the compound indicator L_{den} (day-evening-night noise indicator). According to Article 5 of Directive 2002/49/EC, the noise indicators

L_{den} and L_{nigh} must be used for the calculation of strategic noise maps. L_{den} is derived from L_{day}, $L_{evening}$ and L_{night} using the following formula:

$$L_{den} = 10\lg\frac{1}{24}\left(12*10^{\frac{L_{day}}{10}} + 4*10^{\frac{L_{evening}+5}{10}} + 8*10^{\frac{L_{night}+10}{10}}\right); \quad (1)$$

Considering that noise maps are a powerful zoning and planning resource, investigations of the mean spectrum of noise at each selected location at different times have also been carried out (Pasch et al., 2002). The arguments in favour of its feasibility are given. They show that, in spite of widespread opinion, costs and required time can be reduced considerably by the use of this low-priced method.

Geographic Information Systems (GIS) play an important role in noise mapping. An appropriate use of GIS in mapping noise effects makes it possible to optimise the quality and efficiency of noise effect studies. Furthermore, GIS can influence estimation and exposure of uncertainties. The results of different studies can only be combined or compared if the same indicators for noise exposure and the same assessment methods are used. These effect studies support the decision-making process (Kluijver et al., 2003).

The self-report of noise exposure was compared with the information on noise maps (Heinonen-Guzejev et al., 2000) taking into account measures of self-reported annoyance and noise sensitivity. The data from noise maps is divided into two groups: "exposed" and "not exposed" to transportation noise. The division aims to minimize problems caused by potential misclassification of the data. Finnish authors conclude that self-report of noise-related items can supplement the information on noise maps.

The paper shows both: the practice of road traffic noise mapping according to the EU Directive 2002/49/EC and EC Recommendation C(2003)2807 and noise mapping as a tool for the management of urban traffic flows.

An assessment of street traffic flows

Rapidly growing urban traffic flows have been identified as major contributors to the street traffic noise in a number of medium and large cities. Many cities in the Central Eastern European (CEE) countries have had to face the consequences of a substantial increase in private cars and a crisis of public transport over the last decade (Europe's Environment, 1995). Most cases involved used cars imported from Western Europe into the CEE countries.

The description of the study area

Kaunas, as a medium-size city with 373,669 inhabitants (2003), reflects trends of private car increases and the collapse of public transport in the CEE countries. Most cars are 10-15 years old and are relatively noisy compared with up-to-date equivalents.

The central part of Kaunas , an administrative, commercial and cultural hub, is heavily affected by intensive traffic flows. Traffic jams have become an endemic feature of the city centre due to insufficient street capacity. The majority of the 3-5-storey buildings situated in the city centre aggregate structures directly exposed to traffic noise.

Street traffic flows

The street traffic flows analysis related to noise mapping is based on manual counts, as well as on the analysis and automated traffic flows analysis performed by the Transport and Road Research Institute (TKTI Lithuania, 2001). The study involves measurements of traffic intensity, composition of flows, and driving speed (Nooralahigan et al., 1995).

The analysis of the data sets reveals that the average traffic intensity (ATI), based on 24-hour series, is rather stable during the whole year. Analysis of street traffic flows with respect to driving speed shows that typical speed ranges between 40-60 km/h. The data was registered using Marksman 660 (Golden River Traffic, United Kingdom) counters-classifiers. The integration of the data was performed over 15-minute periods. The stationary counter-classifier in the segment 009-012 (Donelaicio st.) was used to identify the average of 24-hour yearly flow for different vehicle classification groups. Fig. 3 represents the episode distribution of the driving speed in the street segment 009-012.

Fig. 3: The partition of street traffic flows based on the driving speed (K. Donelaicio st.)

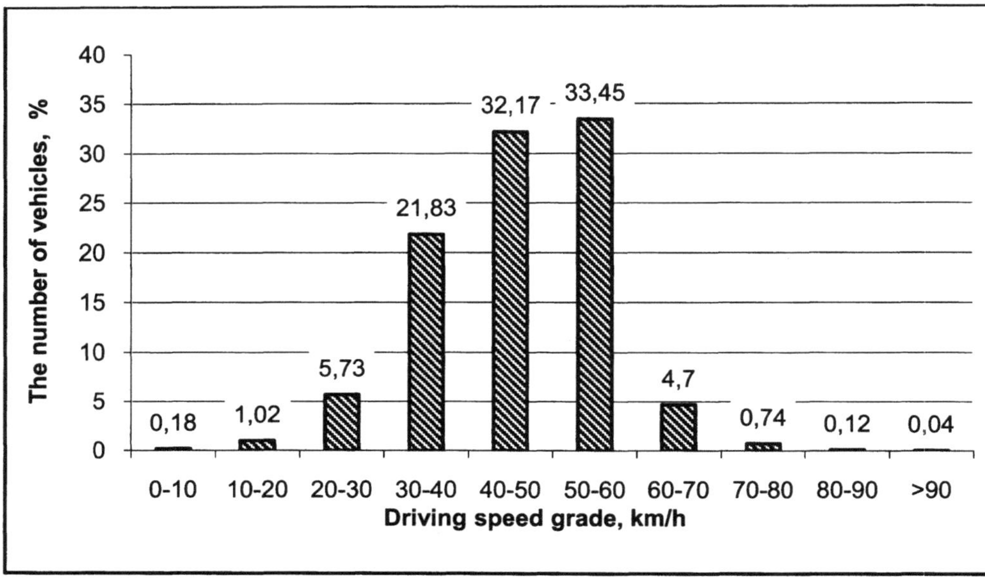

The network of streets (Daganzo, 1997) in the study area is divided into 37 segments, reflecting conditions (the width of street, average speed) typical of the particular segment of the street (see Fig. 4). The specification of flows in each segment is based on the measurement data from 2001. Since data of traffic flows for all street segments within the same year were not available, the authors made a decision to approximate retrospective data sets. In order to determine the corrective factor for the estimated increase in traffic flows from retrospective time series, the authors used comparative data from 37 street segments. As a result, factor 1.027 has been applied to correct data. Traffic flow analysis for the street segments 033-034, 020-022, 010-011, 030-033, 020-030, 035-038, 034-035, 030-031, 030-029 was performed using the approximated data.

Fig. 4: The segmentation of the street grid in the study area

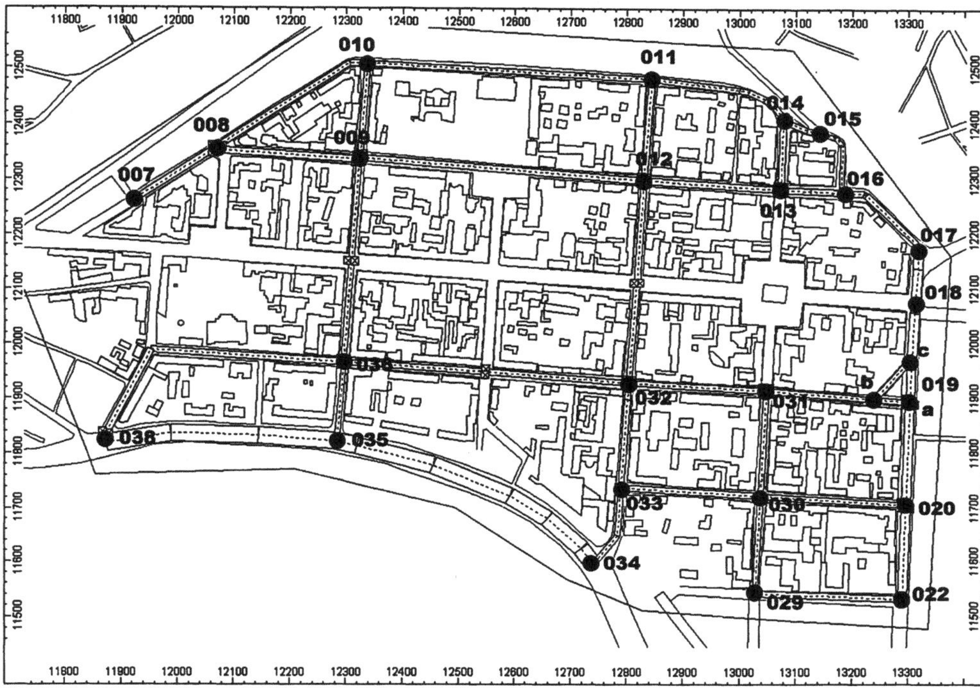

The periodical measurements of traffic flow intensity

Periodical week-long measurements were carried out once each month from June to November. The measurements estimated: a) the annual averages of traffic intensity during 24 hours, b) the variation of traffic intensity averages during 6 months, c) 24-hour traffic intensity for each day of the week, d) features of traffic intensity fluctuations during 24 hours, e) the average speed of traffic flow.

Table 1 presents the average of monthly and yearly 24-hour traffic intensity and ratios between monthly and yearly averages. The averages of weekly measurements served as monthly averages of appropriate months. The authors have noticed that the "commuting circle" reflects temporal variations of traffic flows fairly well and is rather stable over the whole year. It was taken into consideration that meteorological conditions in winter and early spring exert a significant influence on urban traffic flows and show strong correlation with the intensity of flows on out-of-town roads. As a result, the averages of monthly urban traffic flows from January to April were estimated in proportion to the intensity of flows on out of town roads.

Table 1: The average of monthly and yearly 24-hour traffic intensity in Donelaicio st. (street segment 009-012)

	24 hours monthly average (based on daily average traffic counts) and ratios between monthly and yearly averages (%)					
	Month					
	1	2	3	4	5	6
Yearly 24 hours average	9900	10600	11600	12900	13700	13700
	0.77	0.83	0.91	1.01	1.07	1.07
	Month					
	7	8	9	10	11	12
	12600	12200	13900	14100	14200	14200
	0.98	0.95	1.09	1.10	1.11	1.11

Table 2: The average of traffic intensity on weekdays in Donelaicio st. (street segment 009-012)

Measurement period	**Ratio between working and weekend days**	**Weekdays**						
		1	2	3	4	5	6	7
		24 hours yearly averages (vehicles / 24hours) and their ratios with weekly averages (%)						
July-August	1.86	14200	14400	14100	14200	14400	9200	6200
		1.149	1.167	1.139	1.144	1.162	0.740	0.500
September-November	1.57	15300	15400	15600	15600	16500	12100	7900
		1.093	1.094	1.110	1.110	1.173	0.859	0.561

Table 2 presents the average intensity of traffic flows on weekdays respectively for two periods: July-August and September-November. The maximum values of traffic flows were recorded on Fridays. The differences in the intensities of traffic flows from Monday to Thursday ranged within limits of 1.2 per cent. The minimum values of traffic flows were recorded on Sundays and comprised 60.7 per cent of week averages. The values of traffic intensity values on working days exceeded the values of weekends by factor 1.44-1.67. This ratio was even more extensive from July to August.

An estimation of street traffic noise

Noise measurements

Field measurements were performed on working days in the warm period (April-May, 2005) using a tripod mounted Larson Davis (USA), type 4426 equipped with a condenser microphone adapted for free-field conditions and configured for A-weighted frequency response and fast sampling rate (Larson Davis, 2000).

The equivalent continuous sound level L_{eq} was integrated over periods of 5 minutes, which was considered sufficient to average the fluctuating noise from street traffic. The results of the measurement from 10 immission points were used to verify the noise dispersion model for the central part of Kaunas. The majority of the measurements were conducted on working days in the close proximity of the streets, indicated L_{eq} exceeding the 60 dB(A) threshold. The analysis of fluctuations on weekdays shows high correlation (r=0.89) between noise levels and intensity of traffic flows. Instantaneous values of 71.3 dB(A) were reached and were repeatedly registered on Fridays.

The description of the model

The noise dispersion modelling was carried out using CadnaA (DataCustik, Germany) software (Cadna A, 1998) and following the guidelines of the German Standard RLS 90 (RLS-90, 1990). In semi-empirical models, a road is divided into point sources, which are both stationary and continuous. The noise emissions level at each point is relative to the length of the line segment it represents. The number of point noise sources required to properly represent a line source is determined by the distance criterion (Wetzel et al., 1999):

$$l \leq \beta s; \tag{2}$$

with l – the maximum length of each line segment,
β – a constant < 1,
s – the distance from the centre of the segment to the receiving point.

In short, the number of required point noise sources increases as the distance source to receiver decreases. According to the standard RLS 90 guidelines, the value of β was accepted as 0.5. The road traffic source simulates a street with multiple tracks by assuming one-line sources at 0.5 m above the middle of each of the two outer tracks. The equivalent sound level is then computed as the energetic sum of the sources. The model calculates emission levels from relevant traffic flows and associated portion of trucks, vehicle speed, and road surface. In accordance with the software requirements, motor vehicles were classified into light vehicles (less than 3.5 tonnes) and heavy duty vehicles (more than 3.5 tonnes). The following datasets are used as inputs to the noise dispersion model: roads, topography, buildings, reference points for noise level data verification, and grid for data presentation.

The dataset on a carriageway integrated street type and surface, the average number of vehicles per one hour vehicle (vehicles/24), speed limits for light and heavy duty vehicles, uphill and downhill slopes (%), and distance from the street (m).

The datasets on topography and height of constructions were composed from existing GIS maps 1: 5000 (municipality enterprise "Kauno planas") and visualized using CadnaA software. The longitudinal profile was defined as a horizontal carriageway whose gradient in direction of traffic flow is less than 2 per cent. The single-storey houses were attributed 3.4 m heights; respectively every additional storey was given extra 3.4 m height. Due to the fact that the majority of the 3-5-storey buildings in close proximity to the streets make up urban structures exposed to the traffic noise, modelling has included one reflection from the buildings. At the same time, acoustical permeability was equated 0. The 5 m resolution square grid was adopted to present the results of the modelling. The output results were modelled at 4 m above the ground level.

Mapping noise effects

The initial noise level mapping (scenario 0) was based on the analysis of recent traffic flows with respect to the speed limit of 50 km/h. Only in the street segment 038-035-034 (King Mindaugas ave.) was the speed limit 60 km/h. The regressive data analysis was used to verify the model; respectively the linear regression equation $y = 0{,}8319x + 11{,}436$ was used to correct the output data. The validation of the map data shows high correlation ($r = 0.94$) between the measured and modelled datasets. The average equivalent noise level values in the distance of 3 m from the outside track varied from 65.6 dB(A) (the street segment 032-033-034) to 79.1 dB(A) on King Mindaugas Avenue.

Fig. 5: The fragment of the street traffic noise map (speed limit 60km/h)

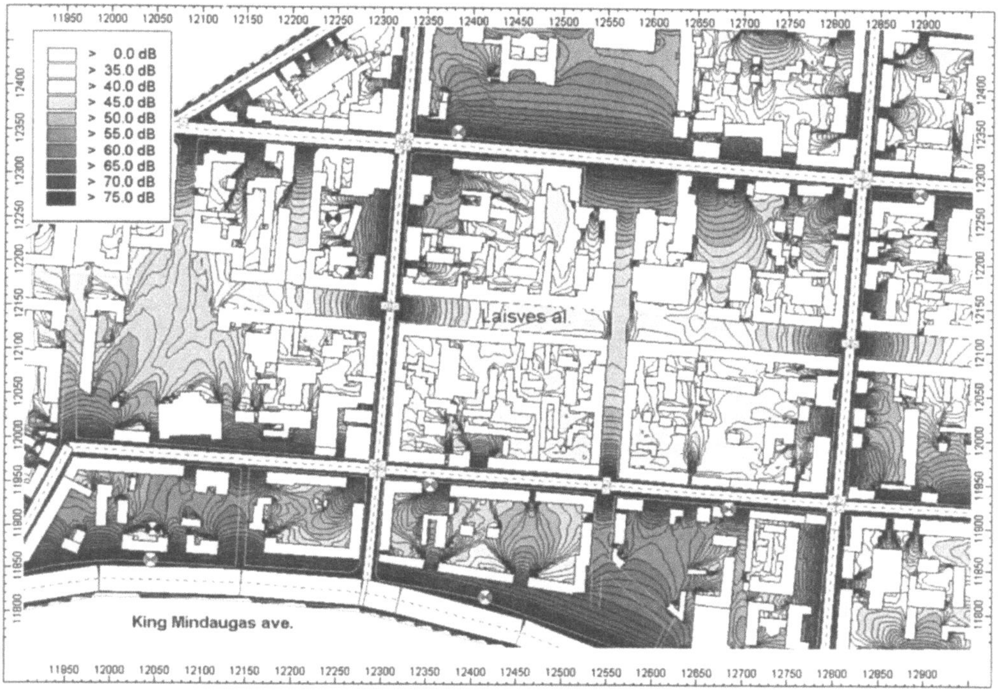

The predictive mapping involved the same composition of traffic flows (Scenario I);, however, the speed limit was raised to 60 km/h (see Fig. 5). Generally speaking, the immission points (Scenario I) exposed to traffic noise were affected by 1.1 dB(A) higher equivalent noise level; L_{eq} from the buildings on the opposite side increased by 1.2 dB(A) in comparison to the initial noise map (Scenario 0).

The predictive modelling with re-distributed traffic flows (scenario II) was performed in the context of *Strategic Traffic System Plan of Kaunas city* (municipality enterprise "Kauno planas"). At the planning level it was considered that integrated networks of public transport would be dominant in the central part of the city (Kliucininkas, 2003). The modelled scenario eliminated 90 per cent of light vehicles (sized up within Scenario 0), replacing them with buses as the alternative mode of transport. The authors came to the conclusion that the type of traffic flow described in the method "XPS 31-133" corresponds to *fluid continuous flow.* The speed limit in the central part of the city was considered to be 50 km/h, except on the King Mindaugas Avenue, where the speed limit v_{max} = 60 km/h. Fig. 6 presents a fragment of noise mapping (Scenario II) for the central part of Kaunas. Compared to Scenario I, the modelled equivalent noise levels at immission points exposed to direct street traffic on average decreased by 6.8 dB(A), respectively; L_{eq} from the buildings on the opposite side decreased by 1.0 dB(A).

Fig. 6: The fragment of the street traffic noise map (re-distributed traffic flows)

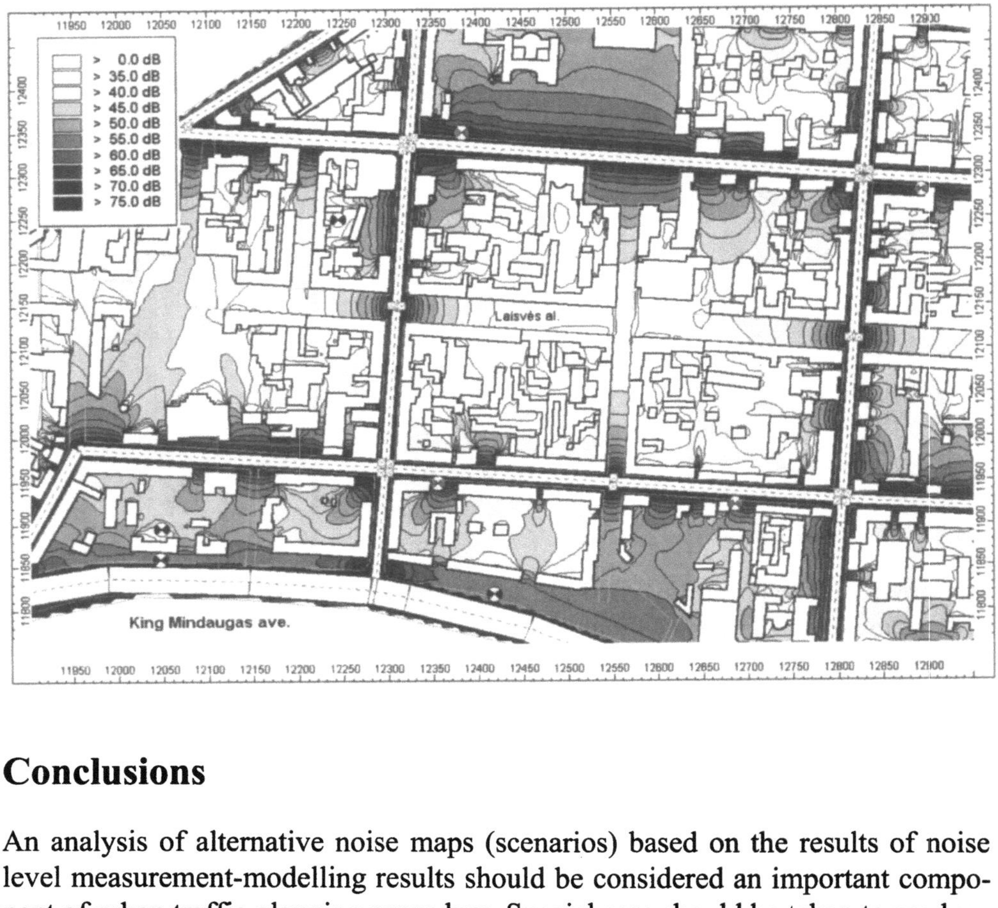

Conclusions

An analysis of alternative noise maps (scenarios) based on the results of noise level measurement-modelling results should be considered an important component of urban traffic planning procedure. Special care should be taken to produce precise datasets for both light and heavy duty vehicles. Predictive modelling with re-distributed traffic flows requires further development of acoustical equivalents for different modes of transport, which has a direct influence on the validation of a model.

This study shows that the description of the traffic flow type gives uncertainty when applied to speed correction. Therefore, the quantitative evaluation of various types of traffic flows is essential to practical applications.

Acknowledgement

The authors would like to express their sincere gratitude to Elena Talockaite from the Environmental Centre for Administration and Technology (ECAT), Lithuania, for her help in mapping noise levels.

References

Cadna, A. (1998). Datacustic GmbH technical documentation and training for immission protection, Munich.

Coelho, B.J. L. (2003). Noise maps and noise action plans for urban noise management. *Proceedings of the tenth international congress on Sound and Vibration,* 5, pp. 2655-2663.

Commission recommendation C(2003)2807 concerning the guidelines on the revised interim computation methods for industrial noise, aircraft noise, road traffic noise and railway noise, and related emission data (2003). *Official Journal of the European Union* L212/49, pp. 49-64.

Daganzo, C.F. (1997). A simple traffic analysis procedure. *Networks and spatial Economics* 1, pp. 77-101.

Directive 2002/49/EC of the European Parliament and of the Council relating to the assessment and management of environmental noise (2002). *Official Journal of the European Communities* L189, pp. 12-25.

European Commission: proposal for a Directive of the European Parliament and of the Council relating to the Assessment and Management of Environmental Noise COM468 (2000).

Europe's Environment (1995). Eds. Stanners, D., Bourdeau, P. ISBN 92-826-5409-5, pp. 359-364. European Environment Agency, Copenhagen.

Heinonen-Guzejev, M., Vourinen, H.S., Kaprio, J., Heikkila, K. and Mussalo-Rauhamaa, H. (2000). Self-report of transportation noise exposure, annoyance and noise sensitivity in relation to noise map information. *Journal of Sound and Vibration* 234 (2), pp. 191-206.

Kliucininkas, L. (2003). Modeling noise effects of urban traffic flows management. *Proceedings of the Tenth International Congress on Sound and Vibration,* 5, pp. 2765-2773.

Kluijver, H. and Stoter, J. (2003). Noise mapping and GIS: optimizing quality and efficiency of noise effect studies. *Computers, Environment and Urban Systems* 27, pp. 85-102.

Larson, Davis (2000). Reference manual. System 824 TM. Precision sound level meter and real time analyzer.

Nooralahigan, A.Y., Kirby, H.R. and McKeown, D. (1995). Vehicle classification by acoustic signature. *Journal of Mathematical Computational Modeling* 27, pp. 205-214.

Pasch, V., Mosconi, P., Yanitelli, M., Cabanellas, S., Vazquez, J. and Rall, J.C. (2002). Spectral information in noise-mapping: An exploratory study. *Proceedings of the First Pan-American/Iberian Meeting on Acoustics*, Cancun, pp. 1213-1220.

RLS-90 (1990). *Richtlinien fur den Larmschutz an Strasen*. Bonn: Der Bundesminister fur Verkehr, Abteilung Strasenbau.

Transporto priemoniu eismo intensyvumo skaiciavimas automatiniais eismo intensyvumo skaitikliais-klasifikatoriais. [Traffic intensity measurements using automatic counters-classifiers] (2001). Report. Kaunas municipality Environmental Protection Department, p. 93 [in Lithuanian].

Wetzel, E., Nicolas, J., Andre, Ph. and Boreux, J.J. (1999). Modelling the propagation pathway of street-traffic noise: practical comparison of German guidelines and real-world measurements, *Journal of Applied Acoustics* 57, pp. 97-107.

IV. Managing urban transportation

4.1 Strategies to improve urban transportation

Commonly observed elements of an urban transport strategy include reduction of car use, improvement in public transport, improvement in the performance of other modes, improvements in the performance of the road network,; and improvement in the performance of vehicles (Nakamura et al., 2004).

Reducing the need to travel. Demands on the transport system can be reduced by enabling people to make fewer journeys, or shorter ones, or both. While reducing the number of journeys may reduce economic and social activity, some journeys can be replaced by telecommunication, as an alternative to travel.

Reducing car use. Car use ensures greater pollution, noise, congestion and accidents per person-km than other modes of transport. Reduction of car use should lead to a reduction in overall traffic levels, and hence to improvements in efficiency, the environment, liveable streets and, if the alternative modes are not significantly less safe, will improve safety.

Improving public transport. Improvements in the coverage and quality of public transport may provide the alternative to car use which enables it to be reduced. Equally, it can provide an enhanced form of transport for those who do not have cars available.

Improving other modes. The same argument holds for walking and cycling, which are the two modes which are available to everyone who does not have a mobility handicap. They also have the advantage of being non-polluting and consuming no energy. Each is limited in its range, but a large proportion of car journeys are over distances which could be travelled on foot or by bicycle.

It is unlikely that improvements to freight traffic would lead to any modal change; however, enhanced logistics could reduce freight movement.

Improving the road network. Whether or not the need to travel, especially by car, is reduced, there will be a need for remaining travellers to use the road network, and there are clear arguments for enabling them to do so efficiently. Improvements can include increases in capacity to allow the management demand for road use to be met more effectively, reallocation of road space between competing uses, and reduction of road space in areas which would most benefit the environment.

Improving the performance of vehicles. Whatever the level of use of cars and public transport vehicles, there is a case for making them safer, quieter, less polluting and more fuel-efficient. Technology instruments introduce improved vehicles designs, which include new motive power, improved emission, noise and safety standards.

4.2 Policies towards sustainable urban transportation

There is very wide range of policy instruments available to those responsible for developing transport strategies for urban areas. One recent study identified some 60 types of instrument, and this excluded those relating specifically to vehicle technology (May and Matthews, 2001). Not surprisingly, there are several ways in which they have been classified, and several general terms; for example travel demand management (VTPI, 2002), which is used loosely to cover wider groups of instruments.

May and co-authors (see Nakamura et al., 2004) distinguish between the way in which instruments work and identify four categories, each with a number of subcategories.

Technology instruments introduce new infrastructure and new vehicle design.

Regulation includes the formal regulation of who may use the transport network and under what terms, and the regulation of transport operators to determine the services they offer and the quality they achieve. But it also includes the less formal management of the transport network and the unregulated decisions of operators on the service they will provide.

Information involves the provision of information to potential and actual users of the transport

network to improve their use of the system or to encourage them to use it in different ways.

Economic instruments use pricing and taxation to influence use and to raise revenue for investment.

Each of these categories of instruments can contribute to the strategies which were identified earlier. Table 10 provides some examples of policy instruments under each category which might contribute to a given strategy. However, it should be stressed that there are few unique relationships here: any one instrument is likely to be able to form part of several strategies.

Table 10: Classification of policy instruments and their contribution to transport and environmental strategies

		Strategies				
		Reduce need to travel	**Reduce car use**	**Improve alternative modes**	**Improve road network**	**Improve vehicles and fuels**
Instruments	**Technology:** *Infrastructure* *Vehicles/Fuel*	– Transit-oriented development …	– Community roads	– Rail and bus infrastructure – New public transport vehicles	– New roads – New parking facilities	– Low Emission Vehicles – Zero Emission Vehicles – Alternative fuels
	Regulation: *Management control service*	– Land use regulations – Suburbanisation control	– Access permits – Parking restrictions – Traffic calming	– Bus priorities – Service improvement …	-Traffic management – Urban traffic control	– Emission regulations – Restriction of low-quality fuel – Vehicle inspection system
	Information: *Advice* *Awareness* *Communication*	– Teleworking	– Awareness campaigns …	– Real time public transport information …	– Driver route guidance – Safety guidance – Traffic information provision …	– Eco-consciousness …
	Economic: *Charges* *Taxation*	– Land taxes …	– Road pricing – Fuel taxes – Vehicle taxes	– Fares policy …	– Road pricing	– Fuel taxes – Green taxes …

4.3 Instruments for urban transportation planning

The information communication technologies (ICT) provide exciting opportunities for urban traffic planners and researchers (Limoges et al., 2000). The areas which are likely to face rapid development in the near future are:

- Intelligent transportation systems;
- Automated data collection;
- Spatial structures and mobility;
- Geographic Information Systems;
- Decision Support Systems and scenario building.

Data collected and archived by *intelligent transportation systems* (ITS) offers great potential in expanding the scope and types of analyses conducted in transportation planning. Archived ITS data is already being used to a great extent for transportation planning, operations, and research. These activities include congestion monitoring, origin-destination studies, planning and traffic model development and validation, and traffic flow research, to name only a few. Given the potential of ITS, there are numerous opportunities as well as challenges in this rapidly growing area.

The continuous 24-hours-a-day, 7-days-a-week fine-grained detail typical of ITS data will eventually supplement or replace traditional traffic data sampling programmes in many urban areas, resulting not only in cost savings but in a dramatic increase in the amount of data available for analysis. Additionally, the large amount and type of data available through ITS will improve existing analyses and enable new types of transportation analyses, such as those focusing on travel reliability or actual discrete choices made by (anonymous) individual travellers or shippers. Numerous data user groups will be brought together because of their interest in obtaining archived ITS data.

The automation of urban data collection offers some significant improvements in data quality by eliminating transcription errors and by introducing online error checking and improved form flow. At the same time, these new technologies eliminate the expense of transcribing written documents into electronic format. Automation further provides for the capture of once burdensome details about urban travel. For example, Global Positioning System (GPS) technology automatically records trip origin and destination, start and finish times, length and duration, and all travel routes. Finally, automation provides a means to capture urban travel data previously unattainable in a survey setting. On-board engine monitors and other vehicle sensors, when installed in personal vehicles, can provide much driver-behaviour-related data, including constant information for vehicle speed, engine speed, throttle position, and engine load.

Spatial policy plays an important part in tackling the issues of mobility. Policies in this area are intended to achieve better spatial coordination of homes, places of work and other facilities. In this way the need to travel can be reduced and growth in car traffic lessened. The intention is also to reduce energy consumption. Although great importance is attached to structural spatial policy, only a fragmentary view of the effects of this policy can be seen: assumptions about mobility impact on land use (particularly the idea that greater urban density leads to reduced car use and thus less energy consumption) are generally correct, but that the effect is not particularly great.

Many of these studies are based on highly aggregated spatial units in which different towns and cities are analysed. These studies explain the collective travel patterns of all households in specific areas, some of which also include information about the socio-economic characteristics of the population in the area. A great deal of attention has been devoted to the impact of spatial density and mixed land-use. Recently, more studies have been carried out at an individual level, in which the travel patterns of the inhabitants of a small number of neighbourhoods were compared with each other. These analyses included corrections for personal and household characteristics. It was investigated to what extent certain spatial features of neighbourhoods provide an explanation for mobility. These studies also showed that spatial planning has little impact.

References

Limoges, E., Purvis, C.L., Turner, S., Wigan, M. and Wolf, J. (2000). *Future of Urban Transportation Data*. Millennium Papers of TRB A1D08: Committee on Urban Transportation Data and Information Systems, 8 p.

May, A.D. and Matthews, B. (2001). *Initial Policy assessment*, PROSPECTS Deliverable 4, Leeds, Institute for Transport Studies.

Nakamura, H., Hayashi, Y. and May, A.D. (eds.) (2004). *Urban Transport and the Environment*. Elsevier Science.

Victoria Transport Policy Institute (VTPI) (2002). *The TDM Encyclopedia*. [online]. Available from: http://www.vtpi.org/tdm.

4.4 Case study IV: Modelling of traffic policy measures for ambient air quality management

Source: Kliučininkas, Linas; Balkevičienė, Simona; Mockuvienė, Jolita; Filho, Walter Leal (2008). Experiences in the modelling of traffic policy measures for ambient air quality management in Lithuania // International Journal of Environment and Pollution. Geneva : Inderscience Enterprises. ISSN 0957-4352, vol. 35, no. 1, pp. 13-24.

This study presents the application of the Integrated Transport Effect Modeling System – ITEMS, taking into consideration existing vehicle fleet, traffic flows, and ambient air quality monitoring data. The objective of this study is to estimate the emissions of CO, NOx, and SO_2 released by motor vehicles in relation to predictive traffic policy measures in Kaunas, Lithuania.

This study analyses the extent to which some traffic policies and future trends may influence the ambient air pollution in urban environment. Three traffic policy measures such as: i) increase the price of car-parking ; ii) reduction of car-parking spaces ; iii) introduction of dedicated traffic lines for public transport are considered as the case aims to examine the extent to which they could reduce emissions from motor vehicles. At the same time, each policy measure comprises four scenarios including one which indicates the current situation. The comparative analyses of the influence of different traffic policy measures on vehicle/passenger kilometres travelled and on emissions are presented.

Introduction

Cities are experiencing notably increasing signs of environmental stress in the form of poor air quality. Automobile transport is now an inherent part of our civilisation and, as has happened with much other technological advancement, the negative aspects are becoming more and more pronounced. A vast majority of the urban population is exposed to conditions that exceed air quality guidelines established by the World Health Organization (EEA, 1998). The use of automobiles has strongly increased during the last few decades. The number of passenger kilometres by private car per capita increased by 90 per cent in Western Europe (OECD, 1996). The number of motorised vehicles in the world grew by about 600 million between 1950 and 1990. Of the 675 million motorised vehicles in 1990, approximately 80 per cent were for passenger transport (Adams, 1999). In Western Europe motor vehicles have overtaken industrial processes in the combustion of coal as the major source of air pollutants.

As motor vehicle traffic is predicted to increase considerably, transport-related emissions are also expected to rise, exacerbating air quality problems.

These problems are persisting in spite of the increasing use of abatement measures. The introduction of catalytic converters is not sufficiently effective in urban areas where travel distances are too short to warm up the converter. Moreover, a substantial part of the pollutants originates from heavy vehicles, for which no effective cleaning technology is yet available. Air pollution from traffic, particularly in street canyons as well as in urban areas, is still a major issue for environmental policy.

This continual traffic growth raises concerns over the impact of traffic emissions on human health and urban environmental quality. It fuels the demand for a coherent regulatory framework for the management of traffic, air quality, and emissions on local and regional scales.

Transport policies can have significant impacts on the environment. Solving any problem of transport impacts one should consider the management issues of transport and traffic systems as well. The sustainable growth respecting the environment is one of the main objectives of the Common Transport Policy (Meyer, 2000). The new urban policy emphasising a compact urban structure with mixed use, high urban planning, and ecological qualities, needs to provide mobility in a sustainable way. At least the following classic strategies are necessary for this:

- Traffic avoidance: avoidance of increased distances throughout more compact city structures, mixed use and concentration on central locations within a region.
- Shift in the means of transport: a large portion of car traffic can be shifted to public transport, bicycle and pedestrian traffic.
- Slowing down traffic and planning of street space: danger, accidents and noise pollution can be significantly reduced by restricting vehicle speed.

The transport policy approaches towards travel in urban areas remain particularly challenging and now focus on the necessity of integrated strategies which combine the use of urban land and transport planning, as well as the improvement of public transport systems and pricing measures. The main goal of the EU environment and transport policies is the reduction of external transport costs.

Overview of Decision Support Systems for urban transport policies

For the first time, the EU Framework Directive 96/62/EEC and the first daughter directive introduce the use of modelling in assessment and management of air quality. The directive recognizes that air quality models are valuable tools for the assessment and forecast of air pollution. They are essential tools in the development of action plans to improve air quality, which is the ultimate goal of member states and the local authorities in order to fulfil their obligations under the directives. Models improve the effectiveness of air quality management.

The application of this numerical tool is also important in the traffic management and in the definition of strategies for the air quality management in urban centres, as well as for the development of future traffic scenarios. Another main advantage to be gained from using models in the assessment and management of air quality is that it enhances the ability to map the spatial distribution of pollutant concentrations, which is not possible with spatial coverage of air quality measurements. This gives a chance to relax the measurement requirements (possibility to reduce a number of stations) and thus to produce a more cost-effective and complete air quality assessment. Models are the only available tools if the impact of possible future sources or alternative future emissions on air quality is to be investigated.

Currently, there are a number of emission models that differ in the fleet composition (vehicle categories and age), driving patterns (average speed only or instantaneous speed and acceleration), covered pollutants, and types of emissions (hot, cold, or evaporative). The selection of the adequate model is an important step to reduce final uncertainties on air quality prediction (Hickman et al., 1999). The locations of the air quality assessment are to be places, where the highest concentration occurs to which the population is likely to be exposed. An air quality model should focus not only on the urban background, but also on calculating representative values at street level. At traffic-oriented sites, the assessment point should be representative of the air quality in a surrounding area of at least 200 m^2 (European Community, 1996).

The ITEMS model

The Integrated Transport Effect Modelling System ITEMS consists of various modules that have been designed and tested, resulting in an urban air quality management tool that can provide reliable answers for policy makers and traffic planners. The ITEMS is made out of five models: transport model, economy model, vehicle stock model, energy-emission model, and environment model. These models are interrelated through a central database from which they extract the input data and where they return their results, which can therefore become inputs of the other models. The model is developed as a cooperative effort of Enerdata S.A. and Grenoble municipality, France.

The ITEMS deals with the urban area only. This includes the centre as well as the closer surroundings of a monocentric city of 100,000 to 800,000 inhabitants. This limitation seems to make sense because the most serious traffic problems occur in urban areas and not within regional areas. Metropoles have a multicentric structure that requires a specific modelling approach (ITEMS, 1999).

To keep the modelling system simple and to make it easier to transfer the ITEMS to other urban areas, only three different areas are distinguished: the city centre, the rest of the urban area and the suburbs. No. networks are considered, with the exception the only link between the three zones.

From an environmental point of view, an energy-emission model is the most important, which calculates energy consumption and pollutant emissions due to the traffic of various vehicles. It is an engineering model which takes into account the different technical influences acting upon energy efficiency and emissions of pollutants, for example:

- cold start
- driving behaviour
- age of vehicles
- technologies of vehicles and engines
- legislation on fuels and emissions
- maintenance programme.

An energy-emission model considers the following pollutants: CO_2, CO, HC, NOx, SOx, particulate matter, VOC.

The street traffic flows in Kaunas

Kaunas, the second major city of Lithuania, is situated in the central part of the country. The administrative area of the city makes up 157.15 km^2 and the number of inhabitants is approaching 400,000. The rapidly growing number of individual cars is identified as a major contributor to the ambient air pollution, especially in the central part of the city. The authors divided the city area into three traffic zones according to the estimates of traffic flows and the character of the built-up areas (see Fig.1).

The analysis of the street traffic flows is based on the manual counts carried out during the years 1999 and 2000, as well as the automated traffic flows analysis, performed by the Transport and Road Research Institute during the period June-November 2001 (TKTI, 2001). The study from 2001 involves measurements of traffic intensity, compositions of flows, and vehicle speed.

The analysis of the data sets for 2001 reveals that the average traffic intensity (ATI) for the central part of Kaunas city, based on 24-hour time series with the exceptional case of summer vacations, is rather stable during the whole year. The analysis of the street traffic flows with respect to the driving speed shows that characteristic speed for the central part of the city ranges between 40-60 km/h. The stationary counter-classifier Marksman 660 (Golden River Traffic, United

Kingdom) was used to identify 24-hour yearly flow averages in the different vehicle classification groups.

Fig. 1: Defined traffic zones in Kaunas

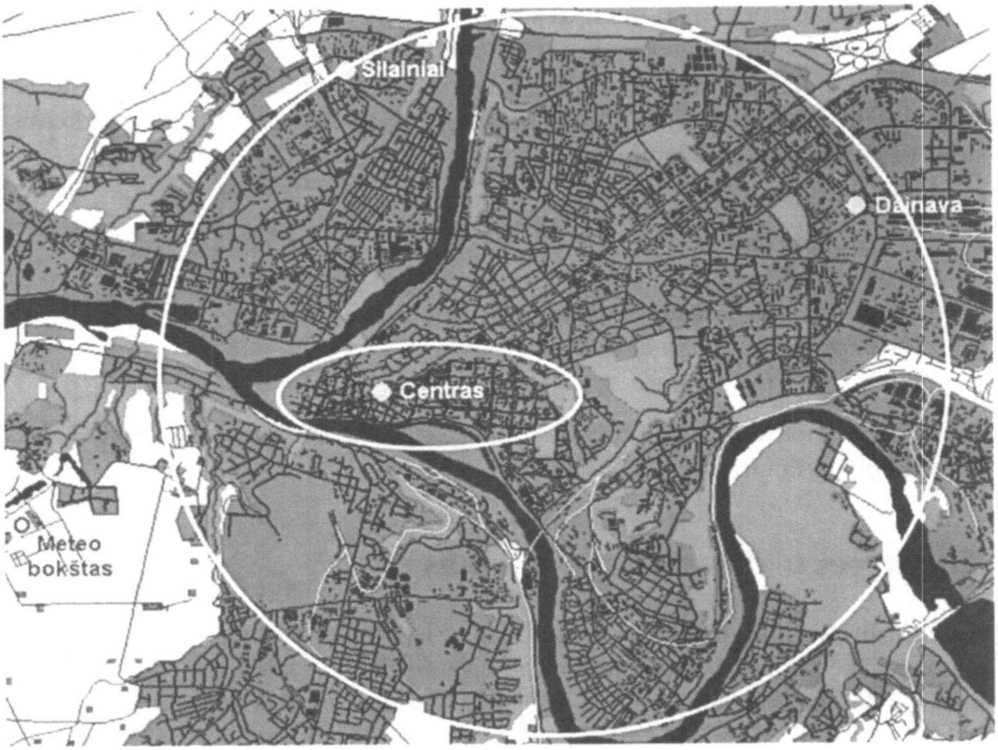

Periodical traffic flow intensity measurements were carried out during one week each month in the period of June-November, 2001. The measurements estimate: a) annual averages of traffic intensity during 24 hours, b) variation of traffic intensity averages during 6 months, c) 24-hour traffic intensity for each day of the week, d) character of traffic intensity fluctuations during 24 hours, e) average speed of traffic flow.

Table 1 presents monthly and yearly 24-hour traffic intensity averages and ratios between monthly and yearly averages. The averages of weekly measurements serve as monthly averages of the appropriate month. The authors believe that "commuting circle" reflects temporal variations of traffic flows fairly well and is rather stable the whole year round. It was taken into account that meteorological conditions during winter and early spring have a significant influence on urban traffic flows and have strong correlation with traffic flows intensity on the out of town roads. As a result, monthly urban traffic flows averages during January-

April are estimated in comparison to the traffic flow intensity on the out-of-town roads.

Table 1: Monthly and yearly 24-hours traffic intensity averages Donelaicio str. (the central part of the city)

Yearly 24-hour average	Month											
	1	2	3	4	5	6	7	8	9	10	11	12
	24-hour monthly average (vehicles/24 hours) and ratios between monthly and yearly averages (%)											
12800	9900	10600	11600	12900	13700	13700	12600	12200	13900	14100	14200	14200
	0.77	0.83	0.91	1.01	1.07	1.07	0.98	0.95	1.09	1.10	1.11	1.11

Table 2 presents the average intensity of traffic flows during weekdays for the periods July-August and September-November respectively. The maximum values of traffic flows were recorded on Fridays. The differences of the traffic flow intensity from Monday to Thursday ranges within limits of 1.2 per cent. The minimum values of the traffic flows were recorded on Sundays and amounted to 60.7 per cent of week averages. The traffic intensity values for working days exceeded weekend values by 1.44-1.67. This ratio was even more extensive during the period July-August.

Modelling transport policy measures

This section examines the extent to which some transport policy measures can affect local levels of pollutants. Each policy measure comprises four scenarios including one which indicates the current situation. This paper presents modelling results for the central part of the city. The analysis is based on the comparison of local air quality from the point of view of different traffic policy measures:

i) increase of car-parking fee
ii) reduction of car-parking space
iii) introduction of dedicated traffic lines for public transport
iv) complex policy measure, considered as combination of the three measures mentioned above.

Table 2: Traffic intensity averages during weekdays, Donelaicio str. (the central part of the city)

Period	**Average 24 hours intensity**			**Ratio between working and week-end days**	**Week days**						
	Total	**Working days**	**Weekend days**		1	2	3	4	5	6	7
					24-hour yearly averages (vehicles/24 hours) and their ratios with weekly averages (%)						
July-August	12400	14300	7700	1.86	14200	14400	14100	14200	14400	9200	6200
					1.149	1.167	1.139	1.144	1.162	0.740	0.500
September-November	14100	15700	10000	1.57	15300	15400	15600	15600	16500	12100	7900
					1.093	1.094	1.110	1.110	1.173	0.859	0.561

Increase of car-parking fee. In the late 1990s, the central part of Kaunas was restricted for private car use. To reduce individual transportation in the central part of the city, the municipality of Kaunas has introduced a 1 Lt parking fee, which corresponds to 0.29 Euros (Scenario a). In order to estimate the extent to which a parking fee could affect transport flows and minimize emissions, the authors have simulated the following hourly parking fees: 2, 3 and 4 Lt (Scenarios b, c, d).

Reduction of car-parking space. The experience of other European cities shows that the reduction of car-parking space can have a significant influence on traffic flows. It reduces transport congestion in the central part of the city; however, it raises parking needs just outside the central zone. The limited number of parking spaces, well-organised public transport, and pedestrian accessibility motivates people travelling into the city centre to moderate their use of private cars.

The authors assume that modelling scenarios will achieve an estimated reduction of 25, 50, 75 per cent in car-parking space (Scenarios b, c, d) in comparison to the current situation (Scenario a).

Introduction of dedicated traffic lines for public transport. Dedicated lines for public transport are applied to areas of intensive traffic flows, in particular to central areas of the cities [10]. It provides opportunities for the fast mobility of public transport and shortens travelling time; however, at the same time it influences travelling-time by car.

In the ITEMS model, this alternative measure is expressed in minutes travelled by public transport or car. It is estimated that, if dedicated transport lines are not introduced in the centre of Kaunas, the same distance will require 7 minutes to travel by car and 17 minutes to travel by public transport (Scenario a). The authors describe several scenarios for the introduction of dedicated traffic lines in the central part of the city:

- shortened travel-time by public transport – 15 per cent, increased travel-time by car – 15 per cent(Scenario b);
- shortened travel-time by public transport – 25 per cent, increased travel-time by car – 25 per cent (Scenario c);
- shortened travel-time by public transport – 35 per cent, increased travel-time by car – 35 per cent(Scenario d).

The validation of the energy-emission model for Kaunas was performed using data series from the stationary monitoring stations *Centras, Dainava,* and *Silainiai* (see Fig. 1). The concentration equation from the linear source was applied to contemporize modelled amounts of emissions with the monitored NOx, CO, SO2 concentrations (Turner, 1964):

$$C(x,y,0:H)=\left(\frac{2q}{\sqrt{2\pi\delta_z u}}\right)\exp\left[-\frac{1}{2}\left(\frac{H}{\delta_z}\right)^2\right] \quad (1)$$

Under the overcast conditions, with an average annual wind speed of 3.37 m/s, correlations δ_z based on the Pasquil stability class D was applied to the equation. The linear source strength q for CO, NOx and SO_2 was calculated and assigned value q_{CO} = 3.47 g/ms, q_{NOx} = 1.98 g/ms, q_{SO2} = 0.022 $*10^{-2}$ g/ms, respectively. The estimation of ambient air concentration was performed at 10 metres distance from the street and the height of the sampling point was 3.5 metres above the ground level.

Results and discussion

The modelling results display changes in total vehicle/passenger kilometres travelled and demonstrate consequent reduction of CO, NOx and SO_2 emissions. The impacts of different traffic policy measures indicate moderation of kilometres travelled by individual car and increase in passenger kilometres travelled by public transport (buses) (Martuzevičius and Kliučininkas, 2000). Each measure comprises four scenarios, described in the previous section. The average changes are shown in Figure 2.

The minor effect is observed after the increase of the parking fee. The taxation has no significant impact on vehicle and passenger kilometres travelled. The "reduction of parking space" provides better results in comparison to the "increase of parking fee"; however, it is less attractive than the "introduction of dedicated lines for public transport". Surprisingly, "the reduction of car-parking spaces" gave a rather limited increase in passenger kilometres. It is most likely that results are affected by a driver's personal behaviour.

The comparison of the individual policy measures has revealed that "the introduction of dedicated lines for public transport" has a major effect on the reduction of emissions from transport. However, the complex implementation of all three measures described will give the most favourable results.

Fig. 2: Estimated traffic flows under different traffic policy measures

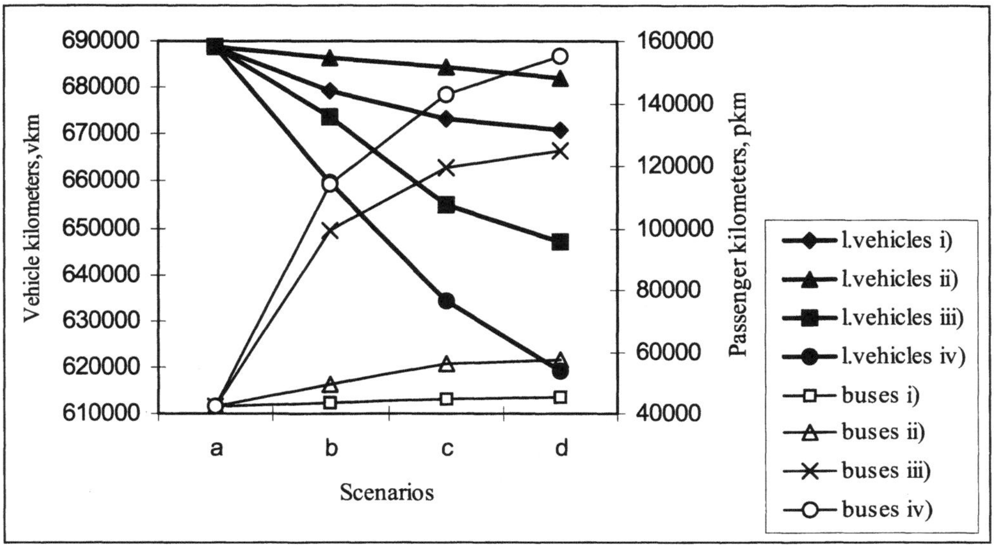

Figs. 3-5 show the effects of changes from different traffic policy measures on average levels of the CO, NOx and SO_2 emissions. The increasing share of passenger kilometres and the decreasing amount of vehicle kilometres produces the evident tendency in the pollution reduction.

Fig. 3: Estimated levels of CO emissions under different traffic policy scenarios

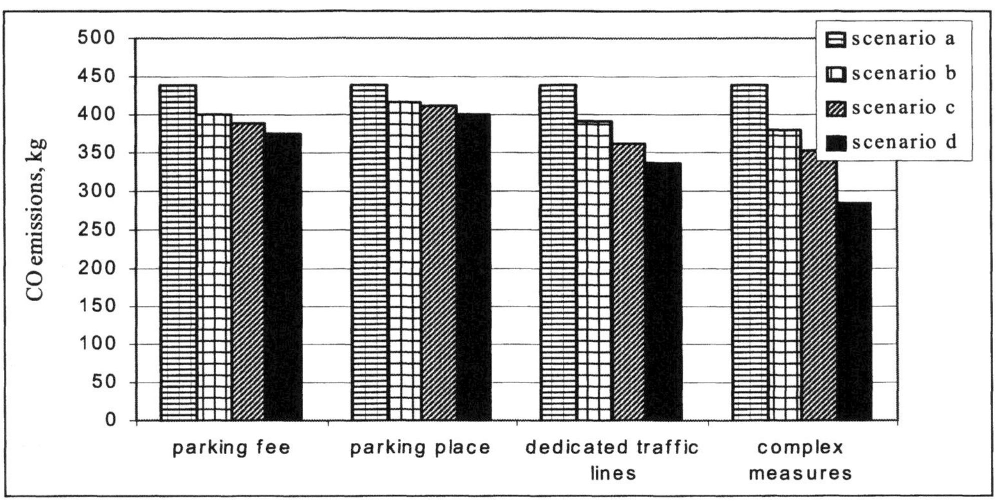

Fig. 4: Estimated levels of NOx emissions under different traffic policy scenarios

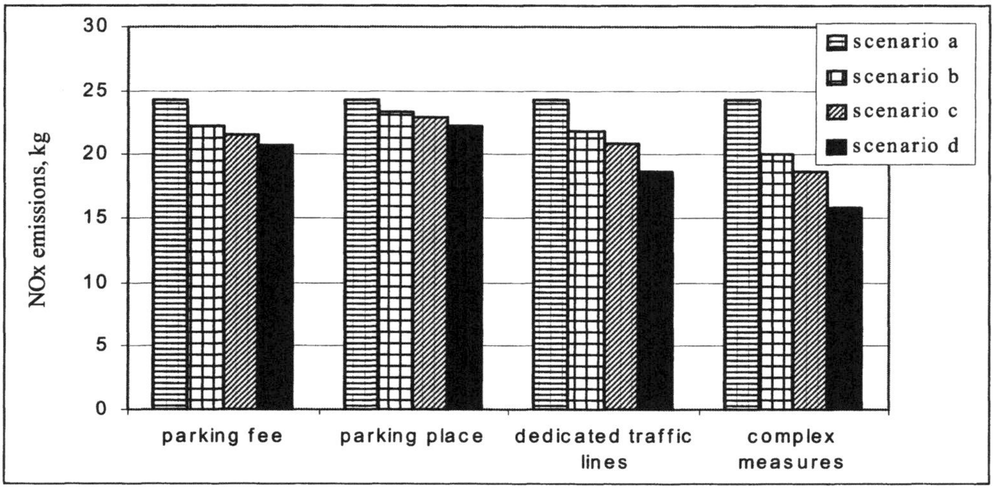

Fig. 5: Estimated levels of SO_2 emissions under different traffic policy scenarios

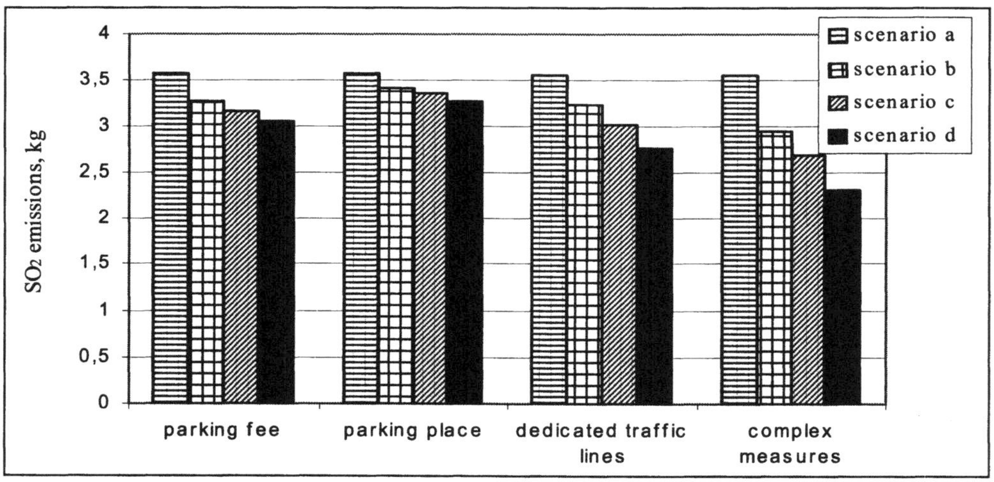

Conclusions

The ITEMS modelling system provides analytical possibilities for the implementation of some traffic policy measures in medium-size cities. The results of the modelling are specific to Kaunas. However, the general trends found may be applicable in other urban areas under similar conditions as well. Since the policy analyses are carried out comparing the results with the "current situation" scenario, it gives "pure" estimates for each policy measure, and eliminates the effects of changes in traffic flows and car fleet.

This paper demonstrates the effects of some transport policy measures and the trends of changes in CO, NOx, and SO_2 emission. The modelling results show that policy measures have varying environmental impacts, particularly for different pollutants. This approach, which provides a range of air pollution indicators from different transport policy measures, is useful in decision-making.

References

Adams, J. (1999). The social implications of hyper mobility. In: *OECD, Project on sustainable transport (EST).* The economic and social implications of sustainable transportation. Proceedings of the Ottawa Workshop, 20-21 October 1998.

Europe's Environment: The Second Assessment (1998). European Environmental Agency, Copenhagen: EEA.

European Community (1996). Council Directive 96/62/EC of 27 September 1996 on ambient air quality assessment and management. *Official Journal of the* EC L 296,0055-0063.

Hickman, J., Hassel, D., Jourmard, R., Samaras, Z. and Sorenson, S. (1999). *Methodology for calculating Transport Emissions and Energy Consumption.* Luxembourg: European Commission, DG VII, 362 pp, ISBN 92-828-6785-4.

ITEMS – Integrated Transport Effect Modelling System (1999). Tutorial. Enerdata s.a. Grenoble-Gieres, France.

Martuzevičius, D. and Kliučininkas, L. (2000). Integrated Modeling of Transport Effects in Kaunas City. Proceedings of the International Conference 'City Environment 2000'. September 28-30, 2000 Kaunas. *Technologija*, pp. 82-88.

Meyer, M.D. (2000). Transport planning for urban areas: a retrospective look and future prospects. *Journal of advanced transportation sector* vol. 34, pp. 143-171.

Towards Sustainable Transportation (1996). Paris: OECD Publications.

Transporto priemonių eismo intensyvumo skaičiavimas automatiniais eismo intensyvumo skaitikliais-klasifikatoriais (2001). [Traffic intensity measurements using automatic counters-classifiers] *Report.* Kaunas municipality Environmental Protection Department, p. 93.

Turner, D.B. (1964). A diffusion model for an urban area. *Journal of Applied Meteorology.*

4.5 Economic Policy Instruments

4.5.1 External effects of urban transportation

Each time a vehicle drives a kilometre it causes wear and tear of the roads, a certain amount of noise, air pollution and so on. As an economist might say, public goods or common property resources such as urban space, clear air and so on are consumed at no cost. It could be said that transport produces negative external effects – externalities, such as pollution, noise and congestion – which are not compensated for in any way. It is necessary therefore to formulate policies to compensate for external effects of transportation.

Such policy instruments can be either physical regulations such as a maximum emission level or a tax (which is a fee for each unit of emission). Other policy instruments are subsidies and tradable emission permits. As a complement to physical and/or tax-based instruments, other measures such as information campaigns are very important.

In contrast to regulated air quality pollutants from passenger cars, characteristics related to the fuel economy are generally regulated as a fleet average or per segment class to allow manufacturers some flexibility in their sales of a range of vehicle sizes. The command and control instrument focuses on the supply side of reducing emissions by requiring the production and sale of fuel-efficient vehicles. In fact, purchasers of fuel-efficient vehicles may even drive more, which is termed the "rebound effect" (Ryan, L., Turton, H. 2007). A rebound effect (also called a Takeback Effect or Offsetting Behaviour) refers to increased consumption that results from actions that increase efficiency and reduce consumer costs (Musters, 1995; VTPI, 2006).

The rebound effect is an extension of the "Law of Demand", a basic principle of economics, which states that, if prices decline, consumption usually increases. These effects may involve reduction in time costs, risk or discomfort. For example, strategies that increase fuel efficiency or reduce traffic congestion, and therefore reduce the per-mile cost of driving, tend to increase total vehicle mileage.

This is not to suggest that the rebound effect eliminates the benefits of efficiency gains. There are usually net congestion reductions or energy savings after the rebound effect occurs. In addition, consumers benefit directly from increased vehicle travel or higher vehicle speeds. However, the rebound effect can significantly change the nature of the benefits that result from a particular policy or project.

There are three types of transport rebound effects defined in the literature:

General traffic: additional vehicle trips on a particular roadway or area that occur when roadway capacity is increased, or travel conditions are improved in other ways. (This may consist of shifts in travel-time, route, mode, destination and frequency)

Induced travel: an increase in total vehicle mileage due to increased motor vehicle trip frequency, longer trip distances or shifts from other modes, but which excludes travel shifted from other times and routes

Latent demand: additional trips that would be made if travel conditions improved (less congested, higher design speeds, lower vehicle costs or tolls)

The study from Victoria Transport Policy Institute (VTPI) reports (Litman, T. 2009) that fuel efficiency standards that increase average vehicle fuel efficiency by 15 per cent are estimated to increase annual vehicle travel by 5 per cent (from 20,000 to 21,000 average annual kms), resulting in 10 per cent net energy savings and a 5 per cent increase in mileage-related costs such as congestion, facility costs and crashes. In this case, the energy conservation, emission reduction and mobility benefits are more than offset by increased mileage-related costs.

Alternative fuels can reduce vehicle operating costs, depending on the type of fuel and its tax rate. This analysis assumes that alternative fuel vehicles are cheaper to drive so that travel increases by 3 per cent, and any additional emissions from the increased mileage is offset by net reductions in per-kilometre emission rates, although actual impacts vary depending on the type of alternative fuel used.

A 15 per cent vehicle fuel price increase would reduce long-term fuel consumption byan estimated 10 per cent, a third of which would be due to reduced vehicle travel. This 3.5 per cent reduction in vehicle kms provides modest reductions in mileage-related costs, resulting in net social benefits, even taking into account the loss of consumer surplus from reduced vehicle travel.

A particular feature that needs to be considered when discussing transport is that there are very many areas of technology involved in transport (different modes of transport, different types of vehicle, engine and fuel) with very big differences in emission characteristics. The negative external effects are, furthermore, quite complex (many types of exhaust chemicals and other effects such as noise, congestion, accidents, barrier effects and so on). The results of these external effects on health and environment are also extremely space and time-dependent. This diversity makes policy more difficult and complex.

Transport policies in most countries rely on regulations, commands and control measures. These include the regulations of traffic flows: traffic zones, parking regulations, minimum standards for vehicles, inspection and maintenance schemes, speed limits and so on.

It is clear that this traditional set of available policy instruments is not sufficient to be able to optimize charges with respect to aspects mentioned above. Many of the external costs vary dramatically and none of these instruments can alone be expected to guide the development of the transport sector.

4.5.2 Eco-efficiency

Eco-efficiency, as a ratio between economic value and environmental impact, may be applied to any unit comprising economic activities, as these activities always relate to cost and values. The units may encompass the whole society, i.e. the macro level. Several options exist for units at micro level of aggregation, i.e. sectors, technologies, product systems, regions and countries.

For the economic part of the eco-efficiency ratio, there are three basic approaches available, all based on life-cycle costing: market cost related values, cost-benefit analysis, and a steady state type of cost, conceptually best linked to steady state models for environmental analysis such as LCA.

For the environmental part, there are fewer consensuses on what constitutes relevant environmental impacts, on the nature of adequate models for their empirical analysis, and on how different types of environmental effects may combine into a single issue. In terms of the modelling of environmental effects, a divergence arises between relatively well-established midpoint empirical modelling, linking emissions and other environmental interventions to environmental effects such as climate change and eutrophication, and more speculative endpoint models, linking environmental interventions to health effects and effects on ecosystems. Due to the long time horizons of many environmental effects it is difficult to take into consideration economical aspect of environmental degradation.

In summary, we can distinguish four main types of eco-efficiency (Huppes, Ishikawa, 2005). The first two are environmental productivity and its inverse, environmental intensity of production, referring to the realm of production. The second pair, environmental improvement cost and its inverse, environmental cost-effectiveness, are defined from an environmental-improvement point of view. The relationship of these variants is shown in Table 11.

Table 11: Four basic variants of eco-efficiency

	Product or production prime	**Environmental improvement prime**
Economy divided by environment	Production/consumption value per unit of environmental impact – **environmental productivity**	Cost per unit of environmental improvement – **environmental improvement cost**
Environment divided by economy	Environmental impact per unit of production/consumption value – **environmental intensity**	Environmental improvement per unit of cost – **environmental cost-effectiveness**

The goal of eco-efficiency is to "decouple resource use and pollutant release from economic activity". This requires breaking the link between the use of nature, as measured by environmental indicators, and economic development as measured by output indicators such as GPD, or transport activity. Decoupling involves a reduction in the negative environmental effects per unit of economic output. This is achieved either by increased efficiency through technological change, or through a shift to a less environmentally damaging product (EEA, 1999).

Indicators of eco-efficiency in the transportation sector are energy consumption and emissions as functions of GDP. An example is passenger cars in Europe, where the average amount of CO2 emission per km travelled by the new passenger cars sold in the EU between 1995 and 2004 has decreased by 12.4 per cent, yet the absolute amount of passenger-car CO2 emissions has still increased owing to the rise in the total number of cars in use.

4.5.3 Maximum abatement costs for calculating cost-effectiveness of green activities

A Maximum Abatement Cost (MAC) method was proposed as a means of preferential purchasing assessment with multiple environmental effects (Oka et al., 2005). The MAC method allows assessment of the cost-effectiveness of introducing a product which emits less of certain pollutants than other more conventional products. In the MAC method, the reduction of a pollutant is multiplied by the MAC, the maximum unit cost of the measures taken elsewhere in society to reduce the pollutant, and is added up over the relevant pollutants. The total sum, called Avoidable Abatement Cost (AAC), is compared with the additional private cost of the product for the purchaser. When the additional private cost is smaller than the AAC, the product is regarded as relatively efficient.

MAC is defined as the highest unit cost, i.e. the cost per kilogram of emission reduction, of the activity that has the highest unit cost among all activities carried out or expected to be carried out soon to reduce the emissions of a substance or a group of substances.

In order to determine the MAC, the activity incurring the highest unit cost must be specified. Strictly speaking, when a particular person or company is carrying out an activity with an exceptionally high unit reduction cost, this very high unit cost may be adopted as the MAC value. Also, when an activity can be divided into several parts that have different unit costs, the partial activity with the highest unit cost may be adopted as the one representing the MAC. However, it is difficult to identify a small activity with very high unit cost contributing only a tiny part of the emission reduction, and it would be appropriate to use the value for an exceptional activity as a reference value for assessing other activities.

One problem with the MAC method is how to allocate the cost of an activity that reduces emissions of several substances. This usually is avoided by identifying activities that predominantly reduce only one substance or one impact category.

Measuring MAC involves assessing the ratio of cost and emission reduction. If the time of incurrence of cost does not coincide with that of the occurrence of the reduction, it is justifiable to account for this difference by time discounting. Thus, if the discount rate changes, the results will also change.

Box 3

MAC related to legally mandated automobile NOx controls

Since existing Japanese regulations failed to meet the environmental standard on nitrogen dioxide (NO_2) in metropolitan areas, new measures to reduce nitrogen oxides in motor vehicle exhaust gas emissions were introduced in 1992 as a "Special Measures Law on Reduction of Total Emissions of Nitrogen Oxides from Automobiles in Designated Areas". This Law was amended in 2001 and was renamed "Special Measures Law on Reduction of Total Emissions of Nitrogen Oxides and Particulate Matter from Automobiles in Specified Areas". The objective of the study was to obtain the MAC for NOx reduction only.

One of the key elements of this law was that for areas where it had been difficult to meet environmental standards for nitrogen dioxide with conventional control measures (the so-called designated areas), emission limits for the vehicles used in these areas (designated vehicles) were set at a stricter level than previous emission limits. The designated areas were the Tokyo and Osaka areas (later, the Aichi and Mie regions were added). Designated vehicles included lorries, buses and specially permitted commercial vehicles based in the designated areas. An emission standard equivalent to the strictest standard for each gross vehicle mass class under the Air Pollution Control Law was applied to these designated vehicles. A key point is that the same standards were applied to both gasoline and diesel vehicles. Another feature of the regulations under this law was that they were applied not only to newly produced or registered vehicles but also to vehicles that were already being used. Designated vehicles that did not meet these standards would not pass the mandatory vehicle inspections and could not be used.

If the price of the vehicles is P, the number of years of shortening is t and the time discount rate is i, the cost is expressed as:

$$P(1-e^{-it}), \qquad (32)$$

based on the estimations of real interest rate for recent years, a 3 per cent discount rate was applied.

The number of years of shortening is given by the average number of remaining service years in the absence of the regulations. For example the vehicle scrapping rate for 0.25-year-old lorries has a value of 0.86 per cent. This means that 0.86 per cent of the standard diesel lorries between 0.25and 1.25 years old were scrapped during the 1-year period. Based on these scrapping rates, the average remaining service life for each vehicle age can be calculated from the following formula:

$$L(t) = \sum_{k=t}^{T} s(k) / s(t)\,. \qquad (33)$$

The same method can be used to calculate the cost of reducing the service life for other types of vehicles.

When designated vehicles not meeting the emissions standards are scrapped, it is anticipated that these would be replaced by vehicles meeting the standard. The present value of the total emission reduction was calculated by the following equation:

$$S_0^T Q e^{-it} dt \equiv Q / i(1 - e^{-it})\,, \qquad (34)$$

here Q is the annual emission reduction per vehicle, i the annual discount rate, T the remaining service life.

The unit reduction costs for all vehicles can be obtained using the ratio of vehicle price, service life and emission reduction, multiplied by the number of scrapped vehicles.

Source: Huppes G., Masanobu I. (eds.) (2007). Quantified Eco-Efficiency: an Introduction with Applications, Springer.

References

EEA (1999). Environment in the European Union at the turn of the Century, Environmental assessment report No. 2, European Environment Agency, Copenhagen.

Huppes, G. and Ishikawa, M. (2005). Eco-efficiency and its terminology. *Journal of Industrial Ecology* 9 (4), pp. 43-46.

Litman, T. (2011). *Efficient Vehicles versus Efficient Transportation.* Comparing Transportation energy Conservation Strategies [online], available at http://www.vtpi.org/cafe.pdf, accesses September 2011.

Musters, A.P.A. (1995). *The Rebound Effect: An Introduction*, Netherlands: Netherlands Energy research Foundation.

Oka, T., Masanobu, I., Yohifumi, Fujii and Gjalt, H. (2008). Calculating Cost-Effectiveness for Activities with Multiple Environmental Effects Using the Maximux Abatement Costs Method. *Industrial Ecology*, 9 (4), pp. 97-103.

Ryan, L. and Turton, H. (2007). *Sustainable Automobile Transportation. Shaping Climate Change Policy*. ESRI Studies Series on the Environment, p. 293.

VTPI (Victoria Transport Policy Institute) (2011). *Transport Demand management Encyclopedia* [online], available at http://www.vtpi.org/tdm/index.php, accessed September 2011.

4.6 Case study V: Comparative analysis of bus and trolleybus related GHG emissions and costs in Lithuania

Source: *Kliučininkas, L. and Matulevičius, J. (2009). Comparative analysis of bus and trolleybus related GHGs emissions and costs in Lithuania // The 5th International Vilnius Conference and EURO-Mini Conference "Knowledge-Based Technologies and OR methodologies for Strategic Decisions of Sustainable Development" (KORSD-2009), September 30 – October 3, 2009, Vilnius, Lithuania / Eds. M. Grasserbauer, L. Sakalauskas, E.K. Zavadskas. Vilnius: Technika, 2009. ISBN 9789955284826. pp. 187-191.*

The study aims to provide a comparative analysis of costs related to emissions from urban bus and trolleybus. The external costs (Euro per km) of greenhouse gases were analysed for the selected buses *Solaris Urbino 12* and trolleybus *Solaris Trollino 12AC*. COPERT 4 software was applied to model greenhouse gases emission factors for the respective buses. Estimation of indirect emissions of greenhouse gases for the trolleybus was based on analysis of primary sources of electric energy production in Lithuania. Monetary values of GHG emissions were derived from the cost factors developed by the HEATCO (Harmonised European Approaches for Transport Costing) project. The study results show that replacement of bus by trolleybus will decrease GHG emissions by 389.69g CO_2/km in 2009 and 287.09g CO_2/km in 2010, at the same time replacement will save 0.042 €/km and 0.039 €/km for each respective year.

Introduction

The main environmental issues in towns and cities are related to the predominance of oil as a transport fuel, which generates greenhouse gases (GHGs) and air pollutant emissions. Usually, the contribution from the transport sector is estimated from the statistical data on fuel consumption or mobility of citizens living within the local authority area. Increased kilometres travelled reflect increased local contribution to global climate change (Kliucininkas and Macionyte, 2008).

CO_2 emissions from new passenger cars sold in the EU decreased by 12.4 per cent between 1995 and 2004, following a voluntary agreement between the European Commission and industry. To enable the EU to reach its 120 g objective by 2012, the Commission, in a Communication of February 2007, outlined a comprehensive new strategy. A legislative framework should ensure 130 g CO2/ km through improvements in vehicle motor technology, and a further reduction of 10 g CO2/ km through other technological improvements and through an increased use of biofuels. Pollutant emissions from vehicles have also been successfully reduced through a gradual tightening of the EURO emissions standards (Green Paper, 2007).

Most of the National Strategic Reference Frameworks submitted by the member states include sustainable urban transport as an area for action. Extension, rehabilitation and upgrading of clean urban public transport such as trolley-buses, trams, metros and suburban rail as well as other sustainable urban transport projects should continue to be promoted and supported by the EU.

Costs related to transport-induced GHG emissions

Different approaches have been adopted to support estimation of costs related to transport-induced greenhouse gases emissions. One way to combine different environmental impacts of energy is to evaluate the so-called external costs or externalities. These are defined as costs that are not included in the price of energy.

Many impacts of energy production on the environment include such costs to society. In principle, society can include (internalise) some of the external costs in the price of energy, e.g. by imposing taxes. In practice, however, it may be difficult to assess the costs, since the knowledge of many environmental impacts is limited and the impacts may occur only after a long delay; the natural values or resources may also have no market values. Life cycle assessment has been applied to comparative evaluation of alternative automotive fuels and technologies which are expected to become available in the near future. At present, 99 per cent of the energy consumed in road transportation is based on crude oil. Carbon dioxide emissions result not only from fuel combustion by the vehicle, but also from fuel extraction, transport, production and distribution.

Alternative fuel chains can involve the use of alternative primary energy sources, innovative fuel production technologies, new automotive fuels or innovative vehicle power-trains. Primary energy sources besides crude oil can be natural gas, biomass, hydro, wind or solar energy. Since there are so many combinations of fuel power-trains, it has been customary to perform the life cycle assessment in two stages. The first stage is called "well-to-tank" and comprises fuel extraction, transport, production and distribution. The second stage is called "tank-to-wheel" and comprises conversion of fuel energy into motion of the vehicle. A complete life cycle assessment combines the results of these two stages and is called "well-to-wheel" (World Energy Council, 2004).

The EU has taken an initiative to develop harmonised guidelines for transport costing. The monetary values of transport emissions were estimated by HEATCO (Harmonised European Approaches for Transport Costing and Project Assessment) project (Heatco, 2006). The monetary valuation was based on the principles of welfare economics, contributing in the long run to consistency in transport costs. The HEATCO project provides cost factors in euro per tonne of pollutant emitted for the majority of EU countries. Valuation of air pollution effects is based on the damages caused by air pollutant emissions. The types of costs included in the valuation approaches are costs for human health impacts, agricultural and forestry production loss, as well as soiling and corrosion of building materials.

The authors of this paper have applied HEATCO monetary values to analyse costs related to GHG emissions from urban bus and trolleybus.

Estimating greenhouse gas emissions

The first step in this study was to calculate emission factors of greenhouse gases for bus and estimate indirect emissions related to trolleybus kilometres of travel. Because of its wide acceptance within the European Union and specialization in the estimation of pollutant emission, COPERT 4 software was selected for this task (Copert 4, 2008).

COPERT 4 is an MS Windows software program designed to calculate of air pollutant emissions from road transport. The development of COPERT has been financed by the European Environment Agency (EEA), in the framework of the activities of the European Topic Centre on Air and Climate Change. The COPERT 4 methodology is also part of the EMEP/CORINAIR Emission Inventory Guidebook. It is the most commonly used model in Europe for official national inventories of emissions from road traffic.

Bus Solaris Urbino 12

This type of bus falls into the "midi urban bus" category, known as HD Euro III – 2000. One hundred per cent of the driving of the bus was in urban conditions. The average speed of the bus was accepted as 30 km/h. Emission factors were estimated for diesel buses with Euro 3 (2000) 98/69/EC emissions standard. To model emission factors, the following input data for the city bus Solaris Urbino 12 were used: sulphur quantity in diesel fuel – 10 mg/kg (diesel fuel density – 845 kg/m^3; PAHs by vol. – 11%; cetane number – 51.95 %, distillation temperature (T95) – 360 °C). It is very important to assess the slope class of road. The selection of the slope class influences emissions and fuel consumption. Emissions and fuel consumption strongly increase when the bus climbs a sloping road. The emission factors (g/km) of GHGs emitted by the bus are provided in the Table 1.

Table 1: Emissions factors (g/km) of GHGs from the bus Solaris Urbino 12

Emissions factors, g/km	Road slope class (%)			
	0%	2%	4%	6%
CO_2	723.05	1074.45	1459.69	1857.98
CH_4	0.0977	0.0977	0.0977	0.0977
N_2O	0.0057	0.0057	0.0057	0.0057
CO_2 equivalent*	727.19	1078.59	1463.83	1862.12

** Assessment is based on Global Warming Potential (GWP)*

Trolleybus Solaris Trollino 12AC

Like other electric vehicles, the trolleybus does not release pollutant emissions directly. However, emissions are produced during generation of electricity at centralised power plants. Emissions from generation of electricity depend on the method of generation, plant characteristics (i.e. type, capacity factor, efficiency, technologies of emission control and lifetime) and source of primary energy. Emissions like SO_2 vary due to the levels of sulphur in fossil fuel. The biggest emissions are produced by fossil-fuel-based power plants. Nuclear power and renewable sources, unlike fossil fuel, do not generate emissions directly. For nuclear power and almost all renewable sources, there are no emissions at the point of generation, but there are releases during the mining and processing of the fuel, construction of the plant, disposal of spent fuel and by-products, and waste management and decommissioning. The followings figures present emissions and GHG emissions in g/kWh from electricity production systems (life cycle analysis) (World Energy Council, 2004).

The data of electricity (fuel) consumption (kWh/km) was obtained from the Kaunas Trolley Bus Company. The average electricity (fuel) consumption of Solaris Trollino 12AC is 0.9 kWh per kilometre. One should consider that enlargement of electric power consumption during uphill movement is compensated during downhill movement. Thus, slope class of road has no influence on consumption of electric power.

The GHG emissions depend on the local mix of electric power production system. Currently 70 per cent of the Lithuania's electric power is produced at Ignalina Nuclear Power Plant. The power plant from 2010 is out of date, thus emissions from the electric power production sector in Lithuania have increased.

Fig. 1: GHG emissions in g-CO_2/kWh from electricity production systems (life cycle analysis).

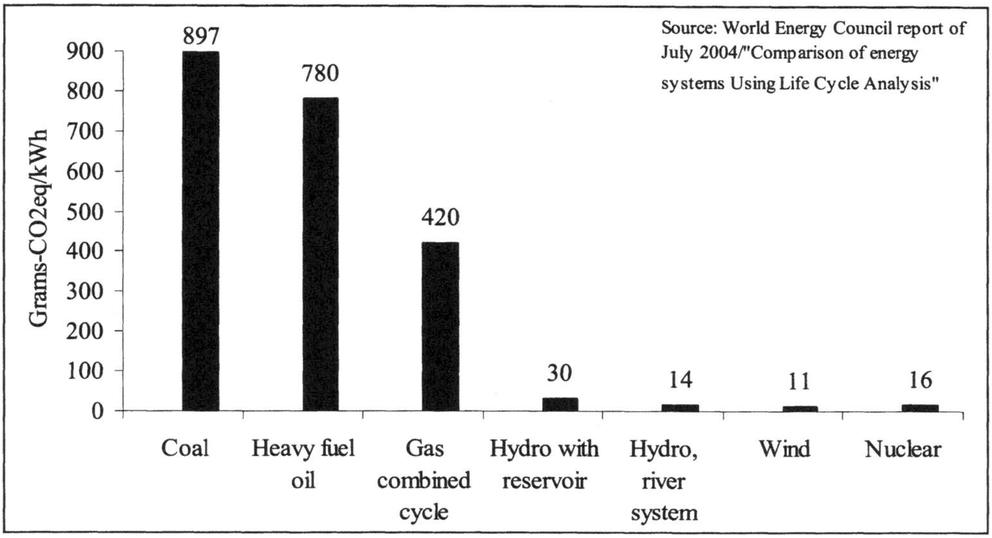

Estimated results of CO_2-equivalent (GHGs) emission factors of trolleybus for the year 2009 and 2010 are given in the Table 2.

Table 2: CO_2-equivalent (GHGs) emission factors of trolleybus

Emissions factors, g/km	**CO_2-equivalent**
Year 2009	337.5
Year 2010	440.1

Calculating monetary values of GHG emissions

The method of calculating costs due to the emission of greenhouse gases (GHG) (usually expressed as CO_2 equivalents) basically consists of multiplying the amount of CO_2 equivalents emitted by a cost factor. The CO_2 equivalent of a greenhouse gas is derived by multiplying the amount of the gas by the associated Global Warming Potential (GWP). The GWP for methane is 25, for nitrous suboxide 298, and for CO_2 it is 1. Carbon dioxide, methane and nitrous suboxide are also called "direct GHGs". Some of the local pollutants – CO, NMVOC, NO_x, and SO_2 – contribute to global warming by forming other species such as ozone and sulphate aerosol particles, and for this reason they are also called "indirect GHGs". Monetary values of indirect GHGs are not estimated in this study.

Due to the global scale of the damage caused, it makes no difference how and where in Europe the emissions of greenhouse gases take place. The value of 22 € per tonne of CO_2-equivalent emitted in 2000-2009 is recommended for all countries. Costs factors of CO_2-equivalent emissions for 2010-2019 are estimated to be 26 €/t.

Values of emissions are calculated using HEATCO factor prices (Bickel and Droste-Franke, 2006). Monetary values of emission from bus and trolleybus for the year 2009 and 2010 are given in Tables 3 and 4.

Table 3: Monetary value of bus-related GHG emissions

Road slope class		0%	2%	4%	6%
Cost, €/km	2009	0.016	0.024	0.032	0.041
	2010	0.019	0.028	0.038	0.048

Table 4: Monetary values of trolleybus-related GHG emissions

Emissions	CO_2-equivalent
Cost for 2009, €/km	0.007
Cost for 2010., €/km	0.011

Conclusions

The study responds to the target of the European Council to reduce EU greenhouse gas emissions by 20 per cent by 2020, and provides environmental financial estimates of clean urban public transport.

A comparative study gives an example of alternative strategies for public urban transport development. The results of the study demonstrate that environmental costs, structured for monetary values of CO_2-equivalent emissions (GHGs) per one kilometre of travel, are much higher for bus than for trolleybus. The comparative analysis shows that the replacement of bus by trolleybus will decrease GHG emissions by 389.69g CO_2/km in 2009 and 287.09g CO_2/km in 2010; at the same time, replacement will save 0.042 €/km and 0.039 €/km for each respective year.

Analysis of emission factors and respective monetary values shows that bus transport in hilly cities makes more significant contributions to climate change if compared with relatively flat urban areas. Bus-related CO_2-equivalent emissions for the 0 per cent road slope class amount to 727.19g CO_2/km; for the 4 per cent road slope class, 1463.83g CO_2/km.

The substantial difference between monetary values of the trolleybus-related emissions for the years 2009 and 2010 is determined by the following factors: i) from 2010, the electric power production system in Lithuania will face radical changes; ii) according to the HEATCO methodology, 2010 is the threshold year for the higher monetary cost factors.

This study applies to development of urban public transport system. The proposed scheme could be adjusted to the local conditions of the individual city and serve as a methodology for sustainable decision support in urban transport planning.

References

Bickel, P. and Droste-Franke, B. (2006). *Derivation of fall-back values for impact and cost factors for airborne pollutants.* Annex D to HEATCO Deliverable 5. Institut für Energiewirtschaft und Rationelle Energieanwendung, Universität Stuttgart.

Copert 4 (2008). Available at: http://lat.eng.auth.gr/copert/

Green Paper (2007). *Towards a new culture for urban mobility.* Brussels, COM(2007) 551, p. 23.

HEATCO (2006). *Harmonised European Approaches for Transport Costing and Project Assessment.* Final technical report, p. 20.

Kliucininkas, L. and Macionyte, E. (2008). *Integrated assessment of damage from urban motorised mobility induced ambient air pollution, in the International Conference "Environmental Engineering":* Selected papers, vol. 3. Vilnius, Technika, pp. 987-992.

World Energy Council (2004). *Comparison of energy systems using life cycle assessment.* A special report, p. 62.

V. Tools for urban transportation management

5.1 Traffic flows and models

Traffic flow type is a complementary parameter to speed, which accommodates acceleration, deceleration, engine load, and pulsated or continuous traffic motion. Four categories of traffic flows are defined below (EEA, 2002), (Commission Recommendations, 2003):

Fluid continuous flow. Vehicles move with nearly constant velocity on the road section of interest. It is "fluid" in that the flow is stable in both space and time for periods of at least ten minutes. Variations during the course of a day may be observed but without abrupt or rhythmic variations. Furthermore, it is neither accelerated nor decelerated but remains at a steady velocity. This flow type corresponds to the traffic on a motorway/autobahn link or an inter-urban road, on an urban expressway (outside rush hours), and on major roads in an urban environment.

Pulsed continuous flow. A flow with a significant proportion of vehicles in a transitory state (i.e. either accelerating or decelerating), which is stable neither in time (i.e. there exist abrupt variations of flow during small time periods), nor in space (i.e. at any given moment in time, irregular concentrations of vehicles exist on the road section of interest). It remains possible, however, to define an average overall velocity for this type of flow which is stable and repetitive for a sufficiently long period of time. This flow type corresponds to that found on city-centre roads, on major roads close to saturation, on dispatching or connecting roads with numerous crossings, in car parks, at pedestrian crossings and at junctions to residential areas.

Pulsed accelerated flow. This is a pulsed flow, and therefore turbulent. However, a significant proportion of all vehicles are accelerating, which in turn means that the notion of speed has a meaning only at discrete points as it is not stable during displacement. This is typically the case for traffic either on expressways after a crossing, at a motorway slip road, at a tollbooth, etc.

Pulsed decelerated flow. This is the opposite of the previous category, in which a significant proportion of vehicles is decelerating. It is generally found on the approach of major urban crossings, on motorways or expressway exits, or on the approach to a tollbooth, etc.

In this chapter, an overview is given of different kinds of traffic models and their characteristics. The term "traffic model" refers to a model that describes road traffic in a network given the travel pattern. The proposed classification is based on the way the different models deal with time and space.

Four categories are distinguished in this chapter:

- Static assignment models,
- Dynamic assignment models,
- Continuum models,
- Micro-simulation models.

5.1.1 Static assignment model

Static assignment is the classic approach in traffic modelling: the first transport models developed were of this type. In the sixties, when computers appeared on the scene, it became possible to calculate flows on networks. The rather simple modelling technique and limited data-requirements made static modelling the logical first step. Transport demand, the number of trips that are made between zones in the study area, is assumed as a given.

Static traffic assignment (Sheffi, Y., 1985) is the process of allocating trips in one or more trip matrices to their routes (paths) in the network, resulting in flows on links. The assignment process is typically used to produce a number of indicators, not just flows. The main objectives of traffic assignment are (Willumsen, L.G., 2000), (Ortuzar, J. de D. en Willumsen, L.G., 1995):

- to obtain good aggregate network measures (e.g. total motorway flows)
- to estimate zone-to-zone travel costs (time)
- to obtain reasonable link flows and to identify heavily congested links
- to estimate the routes used between each origin-destination pair
- to analyse which origin-destination pairs use a particular link or route

Static assignment methods are generally hour-based: the origin-destination matrix contains, for instance, the trips of a peak hour. The capacities of the road network are expressed as number of vehicles per hour. As a result, the flows the assignment estimates are average flows per hour. With a static assignment, it is therefore not possible to get an insight into the development of a peak period. Consequently, these types of assignment are usually applied in long-term policy studies. The basic assumption in assignment is that travellers choose the routes that offer the lowest perceived individual costs. This cost can be divided into travel-time and monetary travel costs. Generally speaking, there are three types of static assignment, all using this assumption: *All-or-nothing assignment (AON)*: the simplest route choice and assignment method. This assumes that there are no congestion effects, and that all drivers consider the same attributes for route choice. They weigh and perceive them in the same way. The absence of congestion means that the link costs are fixed per link, and therefore all the drivers from A to B choose the same route. This route will always be the route with the lowest travel costs from A to B.

User equilibrium assignment (UE): when the calculated traffic flow on a link will approximate or exceed the link capacity, the speeds achieved on the link will decrease.

The result is a larger travel-time on this link: this effect is called *capacity restraint*. Due to the increased travel-time on (a set of) links on a route, the route has possibly no longer minimal travel costs. Other routes can now prove more interesting, unless all drivers change to this route. Capacity restraint can be used as a means for spreading the trips on a network. User equilibrium can be reached when all traffic has chosen routes in such a way that no individual trip-maker can reduce his travel costs by switching routes. The equilibrium is known as *Wardrop's first principle*. As a result, all routes used between A and B have equal travel costs, and all routes not used have greater costs. UE assignment takes congestion effects into account.

Stochastic assignment: a drawback of user equilibrium is that this method assumes total knowledge and equality of the drivers, while in the real world every driver will make slightly different choices based on slightly different perceptions. With stochastic methods this problem can be solved, although the necessity of finding second-best routes introduces a set of new problems. Another feature of stochastic methods is that the result of the calculation will be different in each run. That means that a large number of runs are needed in order to obtain significant results. The popularity of stochastic assignment seems to have decreased over the past years.

A problem in most static model applications is that the modelling of different vehicle classes is handled too simply for noise modelling. For instance, results for lorries and motorcycles (which are important for noise issues) are often not distinguishable. However, the static modelling technique is suitable for different vehicle types, although this generally requires more input data.

There are basically two commonly used time period systems in static modelling: *hour-based models* and *24-hour-based models*. The choice of one of these systems depends mostly on the traffic and transport data available. To make a good hour-based model, considerably detailed data is needed on departure times of travellers in the area. An hour-based model tries to represent, for instance, a specific peak hour. The distribution of traffic over the day has to be known in order to estimate the traffic demand in such a period. For the 24-hour model, on the other hand, it is enough to know the number of trips made per person in the area.

In a 24-hour system, it is not possible to take congestion effects into account. All trips of the day are assigned to the network in one iteration. Only an all-or-nothing assignment is suitable for this. In the hour-based system, it is common to use hour-based link capacities, so a user equilibrium assignment can be carried out. In hour-based models, two types of hours are typically included: peak hour and off-peak hour (midday). Some models have both morning and evening peak hours modelled. Evening and night periods are usually not modelled. This is not necessarily a problem: it is possible to derive a night model from traffic counts. Alternatively, as long as congestion does not occur at night, a simpler approach could be used: the use of day/night conversion factors to estimate the traffic in the evening and night.

The following output is generated by a static assignment model:

- origin-destination matrix with travel-times/distances/costs
- flows on links/routes
- speeds on links (although, in practice, static models are very rarely calibrated on speeds, which means that the predicted speeds may not be reliable enough to be used)
- chosen routes per o/d-pair
- o/d pair's use of a link
- aggregate results, such as:
  - total number of kilometres travelled on the network,
  - total time travelled on the network,
  - both indicators can be divided by link type.

All vehicles on a link are assumed to have generally the same speed, so a speed distribution cannot be derived. Accelerations of vehicles are not modelled either.

Newly built roads, or network measures such as a change in capacity or speed of a road, are easy to bring into the static assignment model. More dynamic measures, such as dynamic traffic management or variable message signs, are not easily included. For pricing measures, assignments with different user classes are available.

Measures that influence transport demand, such as measures stimulating public transport, or spatial planning measures, have to be entered into a transport demand model, which generates an origin-destination matrix. This matrix is then the input for the assignment.

The relatively simple nature of static assignment modelling simultaneously introduces a possible pitfall. Due to the aggregation level of results, and the averaging of indicators, careful attention must be paid to the interpretation of the results, as well as to the limitations of static models and the way they are calibrated. Otherwise, erroneous conclusions and a false sense of accuracy are the result.

As static models are usually calibrated on traffic flows, flows resulting from models are generally adequately accurate. Calibrating with flows and speeds is much more difficult, and this is not common practice. Speeds from static models are therefore likely to be less accurate.

Unlike in the past, static assignment will not automatically be the first model type where new developments are tested and implemented. Static and dynamic models are expected to grow towards each other.

5.1.2 Dynamic traffic assignment

Dynamic traffic assignment (DTA) models were and are developed as an improvement of the traditional static assignment. DTA models can be used to generate forecasts of traffic that illustrate how traffic flows and congestion levels vary over time. Such models can therefore better present actual traffic flows and can evaluate in more detail policy options aimed at a more efficient use of the current road network (Malone, K.M. et al., 2003), (Friesz, T.L. and Bernstein, D., 2000).

DTA models can also be used for prediction and control purposes (e.g. by traffic operators) and for online control. In combination with demand and time-of-day models, DTA models are an alternative for static assignment models in the future. At the moment, DTA models are not very common for practical applications, but they are receiving more and more attention. In general, DTA models use time periods of a couple of hours, mostly including one peak period. But simulations of 24-hour periods are also possible (although this will result in long run times). DTA models focus on the performance over time of the network under study.

Resembling static traffic assignment, dynamic traffic assignment is the process of allocating *overtime* trips in one or more trip matrices to their routes (paths) in the network, resulting in *time varying* flows on links. The trip matrices are normally defined in trips per hour or trips per quarter of an hour. For each hour (or quarter) of the total time period a different trip matrix is fed into the DTA model to model a time-varying demand. DTA in general can produce the same figures as the static assignment does, but with DTA more insight is gained into the dynamics of these figures. In general, DTA models can be divided into two separate sub-models: a dynamic network loading model and a route choice model.

The dynamic network loading (DNL) model defines the way in which traffic moves along links in the network. DTA models can differ in the way in which the traffic is transported over the network, by using different DNL models. These models can, for instance, make use of speed-density functions, define the in and outflow rates of vehicles on a link using node modelling or make use of 2nd order traffic flow models. The most important issue here – and a major distinction with respect to static traffic assignment models – is that these DNL models take into account not only the current, but also the historical development of the traffic on the link. Another important remark is that many DTA models require that the time step is smaller than the shortest free-flow travel-time over a link in the network. Otherwise, traffic can traverse the complete link under consideration when this condition does not hold. DTA models can also be distinguished with respect to the way in which the routing of the traffic is modelled. This can be done by using a fixed route set containing all used routes between the different OD pairs. For each route, the flow proportion between an OD pair is defined. A second method uses split fractions at nodes. With the first method, the routes used between any OD-pair can be easily tracked. With the latter method, this is not always possible.

Within the route choice model, the type of assignment that is carried out is defined. Dynamic versions have been developed for the all-or-nothing, user equilibrium or stochastic assignment static assignment types. Without departure time choice, these assignment types basically do the same as in the static case; only that the routes are now re-assigned on a dynamic basis, e.g. every 15 minutes. For the user equilibrium and stochastic assignments, multiple iterations are required.

During each iteration, all traffic for the complete period under study is assigned. Once all traffic has been assigned, and all vehicles have reached their destinations, the next iteration can start. In the (near) future, DTA models will include realistic congestion spill-back modelling. This requires an accurate geometric description of nodes to be able to realistically model spill-back effects.

There are also DTA models that combine an iterative traffic assignment algorithm with a traffic simulation model. It can be expected that both continuum and micro-simulation models will be extended with accurate route choice models. Joining DTA and simulation models will certainly lead to improved traffic flow description.

The following output is generated by a dynamic traffic assignment model:

- travel-times or costs per O/D-pair per route;
- time-varying route flow proportions;
- time-varying travel-times, flows and speeds on links.

And aggregate results, such as:

- total number of kilometres travelled on the network;
- total time travelled on the network.

All indicators can be generated for specific user classes.

When compared to static models, DTA models give a better representation of the fluctuations in traffic flows and speeds, which improves noise calculations. However, data on acceleration and speed distributions are not generated by DTA models. The models do provide a (time) series of speeds for each link, but these are all average speeds on the complete link (for each time step).

Peaks in speed are therefore not modelled.

In addition to what can be modelled in static models, DTA models can model dynamic traffic management measures (e.g. infrastructure and pricing, spatial planning). It is also possible to model the effects of shifts in departure times. Measures influencing traffic flow characteristics cannot be modelled. DTA models can also be used in online applications, e.g. for travel-time prediction and to support Dynamic Traffic Management.

5.1.3 Continuum models

The efforts of physicists in describing road traffic resulted in continuum traffic models. The first attempts in transfer fluidum modelling techniques were done by well-known personalities from physics research (Lighthill, M.J. and Whitham, G.B., 1955), (Prigogine, I. and Herman, R., 1971).

These techniques have been improved further and extensions towards typical road-based physics have been developed. The evolution in model development has not yet led to numerous commercial software packages. Because of that, there is only a limited use of this type of models in practice. Nevertheless, a lot of traffic research is done within this area and these efforts will certainly lead to renewed attention for this type of model, in theoretical as well as in practical applications.

Within a continuum model, vehicles are not treated as separate entities. The discrete nature of traffic is idealized to a homogeneous fluidum. Within this continuum approach, traffic is described using typical variables from physics: density, intensity (also called the flow or flux) and the average speed.

The basic principle of conservation of mass is translated as conservation of vehicles on a road.

Pressure is the driving force for the particles in gasses and fluids. This mechanism is substituted for traffic. Particles in the traffic stream (vehicle-driver entity) have some kind of intelligence.

The behaviour of vehicles on a road is usually described in an empirical driving behaviour function. This function is then used instead of the pressure equation to have a consistent model of traffic on a crowded road.

The fluidum model is a set of mathematical equations. Sometimes this can be solved analytically, but in practice a numerical scheme is used. This means that a link is divided up into cells of lengths between 10 and 500 metres, and that traffic conditions are calculated in time steps of 0.5 to 10 seconds. This results in a detailed description of traffic conditions in a space-time lattice. The traffic density, the flow and the average speed are then known along the road at each time step.

Continuum models traditionally focus on long, crowded roads. The extension of continuum models to road networks requires the formulation of node rules. Nodes have no physical length and act as flow exchange locations between links. By developing node definitions the model can describe traffic at junctions, off-ramps, on-ramps and traffic lights.

Further extensions focus on typical traffic properties. Traffic is not a real homogeneous medium.

Therefore multiclass models are developed where differences in vehicle properties (e.g. long trucks versus small cars) and driving behaviour (e.g. aggressive versus slow acceleration driving behaviour) are developed. Within multilane models, the lanes are modelled as various parallel tubes with exchange possibilities. Other assumptions on empirical traffic behaviour lead to complex higher order models (where traffic pressure is defined) or kinetic models (where overtaking probabilities are calculated).

Since commercial software packages are rare, the use of this type of models is rather limited. Therefore, it is difficult to say anything about the "standard" output of continuum modelling packages. A growing interest in the development of fast continuum models could be seen. This will result in more software packages and application within traffic control and traffic operations.

5.1.4 Micro-simulation

Traffic micro-simulation attempts to model the progression of individual vehicles through a road network during a specified simulation time period. A micro-simulation model breaks the overall simulation period down into a large number of discrete time steps, and within each time step uses a number of individual algorithms to generate decisions for all vehicles within the network area. The decisions made are then used to update vehicle position, speed and acceleration information. The methodology used for the simulation process in micro-models falls into two main categories:

- Models in which the available road space and vehicles are treated as fundamentally separate units, with road space viewed as a continuum;
- Cellular Automaton (CA) models, where road space is divided into a number of discrete segments, each approximating a vehicle length, which may be occupied or unoccupied at any given time.

The ability to model individual vehicles gives micro-simulation models a number of advantages over traditional, static models. Micro-simulation is typically used to study the effects of short-term traffic management schemes, signal control policies or public transport priority schemes. With care and careful calibration, such models may be used to accurately assess the effects of heavy traffic levels and urban congestion. Advanced models may be used to study the impacts of incidents (e.g. shockwave propagation after traffic disruption), lane changing or weaving behaviour, Intelligent Transport Systems (ITS), dynamic route guidance systems or in-vehicle systems such as Intelligent Speed Adaptation (ISA).

On a typical office PC, the scale of micro-simulation models may range from small regions, spanning several junctions, to entire city areas spanning several hundred junctions and several thousand vehicles at any given time. There have also been a number of highly parallel micro-simulation models developed, capable of modelling entire regions within a country or state on multi-processor systems. In order to reduce running costs, it is a general requirement for the simulation package to run at greater than real time

A given micro-simulation package may contain a vast array of sub-modules, depending on the exact application of the package. However, a number of core components in any micro-simulation model may usually be identified:

- a network editor,
- a traffic-signals database,
- a vehicle/driver database,
- a simulation module,
- an output or visualisation module.

Micro-simulation models may produce output on a variety of levels. A typical micro-simulation may provide output at the micro-, meso- or macro-scopic level, depending on the amount and scope of data aggregation applied.

Micro-scale outputs. Initial vehicle/driver parameters: (e.g. vehicle length, initial speed, initial location, chosen route, driver reaction time) from the vehicle generation algorithm.

Vehicle trajectories: (i.e. vehicle ID, class, position, speed and acceleration in each time step). Usually models offer the ability to filter the trajectory information by vehicle class and sampling methodology (e.g. trajectories from selected links, or vehicles performing a certain manoeuvre) to limit the size of output and traffic signal changes (e.g. exact time of stage changes, possibly triggered by simulated detector systems). Simulated detector output: many micro-simulation models offer the ability to place simulated detectors within their networks. Modelled detector output may be of a similar nature to vehicle trajectory information (i.e. vehicle ID, class, speed and acceleration at the instant of passing the receiver location). For the sake of simplicity, some models assume a detector has no physical dimension, existing only as a point location. Other models may allow detectors to occupy a finite length of road space, allowing the calculation of occupancy measures, such as those used by traffic signal control systems. Some interpolation/aggregation of instantaneous vehicle parameters is usually necessary to generate detector output (e.g. where detector location falls in between a vehicle's location in two consecutive time steps). The specific outputs (e.g. stopping times at specific bus stops for given public transport vehicles from a micro-simulation of public transport priority measures, or vehicle specific pollutant emissions rates for environmental assessments) and distributions of parameters (e.g. vehicle speeds and accelerations) are usually derived from micro-scale parameters.

Meso-scale outputs. These outputs include link or route-based output, such as mean flow, speed and journey times. Usually such results may be output either for the entire simulation period, or for smaller time periods within the main period.

Node or junction-based output, such as turning movements undertaken or mean queue lengths on junction arms within a given period are also considered. Generally, most micro-scale parameters may be aggregated to provide summary information in a form comparable to the output found from static or DTA models, with some additional information variability within the simulation period being available. The exact aggregation methodology used will vary depending on spatial and temporal requirements, and may require substantial interpolation and post-processing.

Macro-scale outputs. These outputs aggregate network parameters, such as total number of vehicles simulated within the given period, total kilometres travelled, total network travel-time, mean network speed, global network emissions rates, etc.

The stochastic nature of micro-simulation models means that it is normal practice to repeat simulation runs, with differing initial seed values. This is done to:

- assess the stability of the model, to ensure there are no breakdowns in flow due to problems with the relevant algorithm;
- assess variability in selected parameters between simulation runs;
- test model sensitivity to select parameters.

The exact outputs from a given micro-simulation package are usually tailored to the exact application of the model. Given that the models deal with the road transport network at its most fundamental level, almost any output is theoretically possible, provided that all desired features are present and adequately modelled within the selected package. Developers and vendors of micro-simulation packages are usually willing to work directly with clients to ensure the quality of any modelling.

The ability to harness vehicle kinematics information directly is of primary interest to noise modelling, allowing the emission of vehicle sound power levels to be calculated in greater spatial and temporal detail than for other types of traffic model. The variety of traffic schemes that may be modelled by micro-simulation also exceeds that for other model types. However, the broad scope and level of detail of such models means that a variety of approaches may exist for the precise methodology used to aggregate emissions levels and assign them to the road space.

5.1.5 Overview of traffic models

The classification and suitability study of traffic models can be summarized in overview tables.

Table 12 shows the capability of traffic models to produce the required input for traffic noise source models. For each of the four classes of models an evaluation is made of their suitability to produce reliable traffic volumes, speeds, speed distributions (within an hour over a link, where the speed differences depend on differences in driver characteristics and on changes in traffic conditions), acceleration and the traffic fleet influence (the influence on traffic flow properties depending on the vehicle composition of the traffic flow). These issues are discussed on link level.

Table 12: Suitability of traffic models to produce detailed output

	Static	Dynamic	Continuum	Micro-simulation
Traffic volumes	+	++	–	+/–
Speeds	+	+	++	++
Speed distributions	–	+	+	++
Acceleration	–	–	+	++
Traffic fleet influence	+/–	+/–	+/–	+

++ available and reliable

The low score of continuum and micro-simulation models with regard to traffic volumes is due to their lack of accurate route choice models. However, some micro-simulation packages already offer an assignment model.

The different types of traffic models can all estimate the impact of a range of traffic flow measures. Within the design of noise maps, typical traffic flow measures can be proposed. An evaluation for each traffic model type is made for these categories of noise-reducing traffic flow measures. It should be noted that many measures heavily influence traffic demand and that therefore their effects should be also assessed with a demand model. Table 13 only gives an indicative overview of the ability of the traffic models to model most of these measures, assuming a good modelling of traffic demand.

Table 13: Capability of traffic models to model traffic flow measures

	Static	Dynamic	Continuum	Micro-simulation
Reducing traffic volumes	+	++	–	+/–
Changing traffic conditions	–	+/–	++	++
Changing traffic composition	+/–	+/–	+/–	+

++ available and reliable
+ available, possibly not reliable
– not available

Several traffic model experts are beginning to observe a growing overlap between the different model types. Static assignment is traditionally the last step of the classical four step transport model. It is assumed that this last step will be extended to a dynamic assignment. Dynamic assignment depends heavily on a dynamic representation of travel-times in a network. The modelling of traffic operations to calculate dynamic travel-times is done by a dynamic network loading model (DNL). Traffic modelling experts expect a growing fusion between the model types, because both continuum and micro-simulation models can be used as DNL for the dynamic network assignment. Almost all consulted traffic model experts stress that modelling daily and annual evolutions is not restricted by the traffic models themselves, but by the input for these models (traffic demand). Therefore it can be concluded that the absence of off-peak periods and daily and annual evolutions in current models is caused by current practice, which does not provide traffic demand data for the periods considered. However, the basic data needed to derive demand for the periods needed for noise modelling (day, evening, night) is available.

The experts believe that currently the static traffic model is applied in roughly all major cities and regions. They expect an increasing shift from static to dynamic assignment models in large area applications. This will probably occur simultaneously with the extension of the modelling period from the peak period to a whole day period. However, traffic models of night periods are not expected to be introduced very fast. Furthermore, the use of micro-simulations to model small – but increasingly large – urban and local situations will grow. Complete coverage of a large city or a region with micro-simulation models remains less likely.

References

Commission recommendation of 6 August 2003 concerning, the guidelines on the revised interim computation methods for industrial noise, aircraft noise, road traffic noise and railway noise, and related emission data 2003/613/EC, Official Journal of the European Union, L212/49-L212/64.

EEA (2002). *Environmental signals – benchmarking the Millennium*, Environmental assessment report No. 9. European Environmental Agency. Copenhagen. Denmark, p. 149.

Friesz, T.L. and Bernstein, D. (2000). Analytical dynamic traffic assignment models, in: *Handbook of transport modelling*, ed. Hensher, D.A. and Button, K.J., Oxford, UK, Elsevier Science Ltd.

Lighthill, M.J. and Whitham, G.B. (1955). On kinematic waves. II A theory of traffic flow on long crowded roads. *Proc. Roy. Soc.,* A. 229, pp. 281-345.

Malone, K.M., Bliemer, M.C.J., Menard, C.H. and Versteegt, H.H. (2003). *Testing INDY for use as the Dynamic Traffic Assignment model in the Dutch National Model*, Delft, TNO Inro report 2003-30, June 2003.

Ortuzar, J. de D. en Willumsen, L.G. (1995). *Modelling Transport.* Second Edition, Wiley.

Prigogine, I. and Herman, R. (1971). *Kinetic theory of vehicular traffic.* American Elsevier Publishing Company, New York.

Sheffi, Y. (1985). *Urban Transportation Networks.* Prentice-Hall, Englewood Cliffs, N.J.

Willumsen, L.G. (2000). In *Handbook of Transport modelling* (ed. Hensher, D.A. and Button, K.J.), Elsevier Science, Oxford UK.

5.2 Vehicle kilometres of travel

Improved urban air quality requires an understanding of motor vehicle usage and the ability to predict likely changes, particularly in vehicle kilometres of travel (VKT), which has been used as a surrogate for vehicular emissions (Lyons et al., 2003). There is a necessity for a simple model of private motorised mobility that predicts VKT to a reliable level, using modest and relatively easily obtained aggregate data at a metropolitan area level. Such a model could form the basis of the assessment of future transportation demand and mobile-source emissions in cities, especially where only rudimentary data are available (Cambridge Systematics Inc., 2003)

Analysis of cities across the globe illustrates a strong relationship between vehicle kilometres of travel and urbanised land area. As VKT can be used as a surrogate for vehicular emissions, this provides a method of estimating urban vehicular pollution. These emission rates are incorporated with simple meteorological forcing into a box model to provide a generalised urban air pollution model for transportation within any city and validated against published data. The results suggest direct air pollution benefits from minimising the outward growth of cities, and add weight to urban policies that emphasise compact city planning principles (Lyons et al., 2003).

Considerable research is being undertaken on the growing proliferation of automobiles and motorcycles and the impact their usage has on the economic, environmental and social sustainability of cities. A significant part of this research involves the collection of data on demography, employment, land-use and transportation infrastructure, modes and patterns (e.g. Newman and Kenworthy, 1989; Kenworthy et al., 1999; Kenworthy and Laube, 2001; United States Department of Transportation, 1999). However, data sets alone are generally not sufficient in themselves in helping decision-makers, unless they can be systematised or modelled to reveal something of the way the urban system functions and how it responds to different factors. A reliable predictive model of private motorised mobility is of great practical use, given the large impact of cars on cities. A key to producing an effective and user-friendly policy model on private mobility is to maximise the model's predictive capacity, while minimising the complexity of the input variables. The basis of the model is the thesis that all cities behave consistently and predictably as urban systems. Despite the passage of time, as well as geographical and cultural differences, all cities have a certain systematic behaviour that is the essence of being an urban area and it is this systematic behaviour that defines urban mobility. Thus it should be possible to quantify urban mobility across urban areas through an understanding of this systematic behaviour. This is in keeping with Dendrinos and Mullally (1985), who noted that, in spite of micro-level complexity, basic insights into urban evolution can be obtained by making relatively few strategically placed macro-level observations of urban growth and form. With improved urban data sets, the time is now ripe to renew the quest for generalisation across urban systems (Longley and Mesev, 2000). The key driving forces of a city's private motorised mobility and through dimensional analysis develops a dimensionally homogeneous equation which explains that mobility. This allows users to predict future values of private motorised mobility in individual cities.

A wide-ranging review of the literature reveals that there are a set of economic, physical planning, transport infrastructure and transport service and usage factors that are considered drivers and indicators of private motorised mobility (e.g. Ingram and Liu, 1997, 1998, 1999; Kenworthy et al., 1999; Holtzclaw, 1994; Newman and Kenworthy, 1989, 1999). These variables have been defined by Kenworthy et al. (1999) and are used on the basis that data is generally readily available or is routinely measured in most urban areas. This data is measured at a macro-level in cities, as compared to micro-level design factors and other matters referred to by Holtzclaw (1994).

These variables (factors) have been chosen on the basis that:

- Population in many ways defines the characteristics and size of a city.
- Metropolitan urbanised land area defines the physical size of a city.
- The total number of jobs in the metropolitan area defines the size of the city's job market, which has a direct linkage to its overall mobility via the various transportation modes workers use in their journey to work.
- The total length of a city's road network is indicative of the extent to which transportation, especially private transportation, is catered for.
- The total length of a city's rail network provides an indicator of the quality and extent of non-road-based transportation, which is free of traffic congestion and which competes directly with the convenience and reliability of private transportation.
- The number of passenger cars, motorcycles and taxis, with their corresponding annual kilometres of travel and occupancy, are the direct indicators of a city's private motorised mobility
- The public transport service and usage indicators are a reflection of factors that help to minimise the reliance of a city's population on private motorised mobility
- The use of non- motorised transport modes of walking and cycling are factors that help to minimise the level of private motorised mobility by catering for

many shorter trips of various purposes that would otherwise be undertaken in cars or other modes of private transport.

Their extent depends on the cumulative effect of a number of complex factors related to density, mixed land-use, environmental quality of the public realm and micro-scale design factors

- The economic indicators reflect the wealth of the city and its capacity to afford private motorised mobility, and the infrastructure expenditure variables measure the level of spending on roads versus public transport. Different levels of spending on infrastructure influence the supply of the respective types of transport infrastructure.

Box 4

Urban mobility

Cameron et al. (2003) defined a non-dimensional private motorised mobility indicator for a city, defined as the ratio of actual total annual private motorised passenger kilometres to the potential total annual private motorised passenger kilometres:

$$\Pi_{mob1} = \frac{(\alpha_k \alpha_0)}{365 \beta_p 50}, \qquad (35)$$

where α_k is the total annual vehicle kilometres of travel by passenger vehicles, α_0 the average passenger vehicle occupancy, β_p the total population of the metropolitan area, and 365 represents the number of days in a year.

An average daily trip length of 50 kilometres has been taken as representing a theoretical maximum daily private passenger vehicle travel per person in any city, based upon an average travel-time budget of one hour per person per day (Schafer and Victor, 1997) and an average speed of 50 km/h for road traffic. The 24-hour average road traffic speed in most cities does not exceed 50 kmh^{-1} (Kenworthy and Laube, 2001) and the overall average travel-time per person per day (one hour) has been shown to be fairly constant for centuries (though travel-times for individuals in any city follow a normal distribution).

A city's public transport service can be seen as exerting an important influence on the level of its private motorised mobility. A dimensionless ratio incorporating a city's public transport kilometres of service is therefore included in the model development to gauge its impact on the city's private motorised mobility. A public transport vehicle kilometres (vkt) of service dimensionless ratio is defined as:

$$\Pi_{psk} = \frac{\mu_{vk}}{108300 \lambda_a} \qquad (36)$$

where μ_{vk} is the total annual public transport vehicle kilometres of service and 108,300 public transport vehicle kilometres of service per urban square kilometre represents the highest public transport kilometres of service density value for any city in the data set, plus an allowance of 25 per cent to represent an estimated maximum (Kenworthy and Laube, 2001) and appropriate normalisation factor.

The private motorised mobility equation was tested and developed by using data sets for 1960, 1970, 1980 and 1990. One of the significant features of this analysis is that, despite an increase in the number of cities over the four decades, and a significant change in the appearance of many of these cities, the functional form of the private motorised mobility equation has remained constant. This implies that the systematic behaviour of a city is does not vary over time. The structural form consistency explained more than 85 per cent of the variance. Such consistency lends further support to the hypothesis that, despite the passage of time and cultural, geographical and social differences, there is an inherent systematic feature of urban areas that influences private motorised mobility. In addition, the Millennium Cities Database for Sustainable Transport (Kenworthy and Laube, 2001), based on 1995 urban data sets, was used to verify the model. This data set has incorporated detailed information for an additional 47 cities besides those used in the primary test. Despite a completely different set of cities and a different time period, the simple model was able to explain the bulk of the variance in private motorised mobility.

This provides a simple relative measure of private motorised mobility for any city; through the application of dimensional analysis, Cameron et al. (2003) were able to show that urban private motorised mobility could be represented as:

$$\Pi_{mob1} = 0.38\left(\Pi_{uf}^{-0.26}\right)\left(\frac{1}{\left(1+\exp\left(-3.4\Pi_{vehicle}\right)\right)}\right)^{3.4} \tag{37}$$

where

$$\Pi_{uf} = \frac{\beta_p}{50000\lambda_a} \tag{38}$$

$$\Pi_{vehicle} = \frac{\alpha_c}{0.85\beta_p} \tag{39}$$

α_c is the total number of cars on register, and λ_a the actual urbanised area (Kenworthy et al., 1999) rather than the total area within a city's territorial boundaries. Π_{uf} has been normalised by recognising that, over their entire metropolitan area, cities do not normally exceed 50,000 persons/km^2 (Kenworthy and Laube, 2001); $\Pi_{vehicle}$ was normalised according to Dargay and Gately (1999), who assumed that ownership saturates at 0.85 vehicle per capita.

Incorporating Eqs. (1),(3) and (4), Eq. (2) can be reorganised to yield urban vkt as:

$$\alpha_k = \left(\frac{115552 \beta_p^{0.74} \lambda_a^{0.26} S}{\alpha_0} \right) \tag{40}$$

with

$$S = \left(\frac{1}{\left(1 + \exp\left(-3.4 \left(\frac{\alpha_c}{0.85 \beta_p} \right) \right) \right)} \right)^{3.4} \tag{41}$$

The term *S*, defined by Eq. 41, is a vehicle saturation factor that changes relatively slowly as a function of the number of cars per head of the population, but nevertheless represents an infrastructure limit on the growth of total vkt. It is a measure of automobile ownership and illustrates the impact of economic growth through higher vehicle ownership in accordance with Ingram and Liu (1998) and Dargay and Gately (1999). Implicit within this model of private vkt is the impact of car ownership saturation, as well as population growth.

Source: Cameron, I., Kenworthy, J.R. and Lyons, T.J. (2003). Understanding and predicting private motorised urban mobility. Transport Research Part D 8, pp. 367-283.

References

Cambridge Systematics Inc. (2003). *Simplified travel demand forecasting for developing countries: phase 1.* Final report. Prepared for US Environmental Protection Agency. Cambridge Systematics Inc., Cambridge, Massachusetts, March 2003, 153 pp.

Cameron, I., Kenworthy, J.R. and Lyons, T.J. (2003). Understanding and predicting private motorised urban mobility. *Transport Research Part D* 8, pp. 367-283.

Dargay, J. and Gately, D. (1999). Income's effect on car and vehicle ownership, worldwide: 1960-2015. *Transportation Research Part A* 33, pp. 101-138.

Dendrinos, D.S. and Mullally, H. (1985). *Urban evolution: studies in the mathematical ecology of cities.* Oxford University Press, London, p. 184.

Holtzclaw, J. (1994). Using residential patterns and transit to decrease auto dependence and costs. *Natural Resources Defence Council*, p. 40.
Ingram, G.K. and Liu, Z. (1997). Motorisation and Non-motorised Transport in Asia. Transport system evolution in China, Japan and Indonesia. *Land Use Policy* 3, pp. 69-84.
– (1998). *Vehicles, Roads, and Road Use: Alternative Empirical Specifications.* Policy Research Working Paper 2036. Word Bank, Washington DC, p. 39.
– (1999). *Determinants of Motorization and Road Provision.* Policy Research Working Paper, World Bank, Washington DC, p. 29.
Kenworthy, J.R. and Laube, F.B. (2001). *Millennium Cities Database for Sustainable Transport.* Internationl Union (Association) for Public Transport (UITP), Brussels and Institute for Sustainability and Technology Policy (ISTP), Perth, CD-ROM database.
Kenworthy, J.R., Laube, F.B., Newman, P.W.G., Barter, P., Raad, T., Poboon, C. and Guia, B. Jr. (1999). *An international sourcebook of automobile dependence in cities, 1960-1990.* University Press of Colorado, Boulder, 704 pp.
Longley, P.A. and Mesev, V. (2000). On the measurement and generalisation of urban form. *Environment and Planning* A 32, pp. 473-488.
Lyons, T.J., Kenworthy, J.R., Moy, C. and dos Santos, F. (2003). An international urban air pollution model for the transportation sector. *Transportation Research Part D* 8, pp. 159-167.
Newman, P.W.G. and Kenworthy, J.R. (1989). *Cities and automobile dependence: an international sourcebook.* Gower Technical, Aldershot, UK, 388 pp.
– (1999). *Sustainability and Cities: Overcoming automobile dependence.* Island Press, Washington DC, p. 442.
Schafer, A. and Victor, D. (1997). The past and future of global mobility. *Scientific American* 2777 (4), pp. 36-39.
United States Department of Transportation (1999). Summary of Travel Trends. 1995 Nationwide Personal Transportation Survey. United States Department of Transportation and Fedral Highway Administration, December, p. 142.

5.3 Geographic Information Systems and Spatial Decision Support Systems

5.3.1 Geographic Information Systems

Geographic Information Systems (GIS) are integrating systems that bring together ideas developed in many areas. These include the fields of geo-sciences, cartography, remote sensing, and computing, to name but a few. Understanding of what Geographic Information Systems represent may be helped by considering the component parts of the term separately.

Geographic. This term is used because GIS tend to deal primarily with "geographic" or "spatial" features. These are objects which can be referenced or related to a specific location in space.

Information. This represents the large volume of data which is usually within a GIS. All real world objects have their own particular set of characteristics or descriptive attributes. This data needs to be stored and managed for all spatial features or interest.

Systems. This term is used to represent the systems approach taken by GIS, whereby complex environments are broken down into their component parts for ease of understanding and handling, but are considered to form an integrated whole.

For a number of reasons, GIS is more difficult to define than might at first be imagined. The various ideas about GIS can be synthesized and presented in the form of four distinct but overlapping views. These can be termed the map, database, spatial analysis and application views.

The map view focuses on cartographic aspects of GIS. Supporters of this view see GIS as map processing or display systems. Topographic and thematic mapping agencies also support the map view and place great emphasis on the ability of GIS to produce high quality maps and charts.

The database view of GIS emphasizes the importance of well-designed and implemented database. A sophisticated database management system is seen as an integral part of a GIS. This view is predominant among members of the GIS community who have a background in computer science.

The third view of GIS emphasises the importance of spatial analysis. This view focuses on analysis and modelling in which GIS is seen more as a spatial information science than a technology. This view looks likely to become the most widely accepted by the GIS community. The fourth view is characterised by a great diversity of applications.

In broad terms, Geographic Information Systems are computer-assisted systems for the acquisition, storage, analysis and display of geographically referenced data.

While general cartographic theory and practice are concerned with the study, preparation, production and utilisation of all existing types of maps and map-related products, the subject matter of thematic cartography is more limited and refers to those products which illustrate only selected spatial themes or topics. In most cases GIS users are dealing with thematic maps. Its aim is the graphic representation of spatial patterns and relationships, insofar as these can be related to geographic space and transformed into cartographic symbols (Kliucininkas, 1994).

5.3.2 Decision Science

Formal analyses of decision-making began with the development of Operations Research in response to demands for rational and analytical decision-making during the Second World War (Moore, 1975). Today, decision analysis is treated in a wide range of disciplines, each with its own techniques and focus of study. Generally, studies in decision-making fall into two areas: *prescriptive* and *descriptive* decision analysis.

Descriptive studies in decision-making have their roots in psychology and sociology and concentrate on the search for reasons why decisions are made in the manner that they are. On the other hand, the prescriptive analysis of decisions emphasises the development, evaluation and application of techniques to facilitate decision-making. These studies rely upon the logic of mathematics and statistics and utilise the concepts of utility and probability to analyse decision problems. The concept of utility relates to the expression of preferences among alternative options, while probability serves to evaluate the likelihood of these preferences being realized.

Traditionally, prescriptive decision analysis has taken the form of either an objective or subjective evaluation of decision criteria. In objective analysis, attempts are made to provide a financial appraisal of decision events by identifying all the potential effects and the magnitude of such impacts based on the market value of events and criteria involved. Advocates of this approach include Dasgupta and Pearce (1972) and others who have written about the cost/benefit analysis technique.

Subjective analysis of decision events, on the other hand, comprises various approaches which share the common purpose of helping decision-makers marshal their thoughts, express consistent judgements and choose rationally (Cochrane and Zeleny, 1973; Coleman, 1971; Keeny and Raiffa, 1976; Voogd, 1983). The techniques adopted in the various approaches incorporate explicit statements of preferences by decision-makers. Such preferences are represented by various quantities, weighting schemes, constraints, goals, utilities and other parameters. They include multi-attribute utility theory (Fishburn, 1974), goal programming (Dyer, 1972; Ignizio, 1985; Lee, 1972), linear programming (Fiering, 1986; Zeleny, 1974) and other optimisation techniques (Nijkamp, 1979).

A decision is a choice between alternatives. The alternatives may represent different courses of action, different hypotheses, different land allocations, and so on. Rational human behaviour involves the evaluation of choice alternatives based on some criteria. Decisions may be characterised as *single* or *multi-objective* in nature, based on either a *single* criterion or *multiple criteria.* While one is occasionally concerned with single criterion problems, most problems approached with a GIS are multi-criteria in nature. Our concern lies with how to combine these criteria to arrive at a composite decision. The techniques of *Multi-Criteria* and *Multi-Objective* evaluation are well developed and used for environmental planning.

While many decisions we make are prompted by a single objective, sometimes we also need to make decisions that satisfy several objectives. These objectives may be complementary or conflicting (Carver, 1991). With complementary or non-conflicting objectives, land areas may satisfy more than one objective. Optimal areas would be those that satisfy both of these objectives to the maximum degree possible. With conflicting objectives, objectives compete for the available land, since it can be used for one or the other, but not both. In cases of complementary objectives, multi-objective decisions can often be achieved through a hierarchical extension of the multi-criteria evaluation process. For example, we might assign a weight to each of the objectives and use these along with the suitability maps to combine them into a single suitability map indicating the degree to which areas meet all of the objectives considered (Voogd, 1983). With conflicting objectives it is sometimes possible to rank the objectives and reach a prioritised solution (Rosenthal, 1985). In these cases, the needs of higher ranked objectives are satisfied before those of lower ranked objectives are dealt with. However, this is often not possible, and the most common solution to conflicting objectives is the development of a compromise solution. Undoubtedly the most commonly employed techniques for resolving conflicting objectives are those involving optimisation of a choice function such as mathematical programming (Fiering, 1986) or goal programming (Ignizio, 1985).

Information is vital to the process of decision-making; however, we recognise that we rarely have perfect information. This leads to uncertainty, of which two sources can be identified: database and decision rule uncertainty. Measurement error is the primary source of such uncertainty. Decision rules typically contain procedures for combining criteria into a single composite index and a statement of how alternatives are to be compared using this index. Decision rule uncertainty arises from the manner in which criteria are combined and evaluated to reach a decision. A very simple form of decision rule uncertainty relates to parameters or thresholds used in the decision rule. A more complex issue is the decision rule itself.

5.3.3 Spatial Decision Support Systems for transportation planning

The context of this chapter is intended to present GIS as a decision support tool, with particular emphasis on the areas of traffic planning and environmental management. A spatial Decision Support System (sDSS) is an interactive, computer-based system designed to support a user or group of users in achieving greater effectiveness in decision-making while solving semi-structured spatial problems. It is designed to assist the spatial planner with guidance in making land-use decisions.

The application of sDSS provides decision-makers and planners with an effective tool. Furthermore, experience has shown that it acts as an excellent vehicle for discussion and the development of consensus. It gives the possibility for better understanding and structuring of complex issues related to the decision. The process of decision-making needs to incorporate the participation of a representative group of individuals who have a direct stake in the outcome of the decisions.

Analysis is invaluable when planning new developments: for example roads, buildings and parks may be compared in various alternative scenarios. It is possible to estimate future traffic densities and connected air pollution. Accessibility of new public facilities, business districts and residential areas is another issue which may be addressed using analytical GIS software. The various alternatives may be compared and the best one chosen. With more advanced GIS, optimal locations may be calculated *a priori* – not comparing predetermined alternatives – and used in planning.

Today, computer-based techniques are becoming increasingly important in urban planning. Local-scale GIS is widely used for planning, extending and maintaining transport/traffic systems. The GIS systems are mostly used to keep track of the networks, as in traditional maps, but they may be used also in a more advanced way, especially for many kinds of spatial analysis (Lukosius, S. and Kliucininkas, L., 2003).

This technology offers extremely significant power in transport modelling. The spread of sDSS use facilitates the efficient and portable spatial data storage, updating and processing. In addition, an sDSS system facilitates model accessibility, database maintenance and updating, and the cartographic display of model results. This can greatly enhance the role of the transport model as a decision-support system in transportation planning and policy development. The interface of network-based equilibrium models with the sDSS platform offers substantial potential for modelling transportation planning, analysis and control. The systems can greatly improve the realistic representation of the multi-modal transportation network, increase the likelihood of database integrity relative to the traditional multi-modal network data model, and ensure effective user-interfaces and the efficient visualisation of network equilibrium solutions. Thus, integration of sDSS into complex computing models would provide transport/traffic planners with new analytical possibilities. For example, it would be possible to present the results of cost-effective analysis in the form of thematic maps.

The operation of the sDSS tool is based on the concept of scenarios. A scenario is defined in terms of an appropriate database that represents the transportation infrastructure of the study area at the starting point as well as a set of user-defined policies whose impacts will modify the database. User interaction with the sDSS environment falls into three different functional groups, accessed via a hierarchical menu system. These are: scenario management, scenario analysis and scenario comparison.

Scenario management provides the user with capabilities to create a new or to retrieve a previously stored scenario, and to define a set of policy measures to be examined. A "reference" or "zero-state" scenario is always present, serving as the basis for the creation of new scenarios. The definition of a new policy measure is carried out through interactive procedures based on sets of hierarchical popup dialogue boxes which allow the user to specify the policy type, the geographical objects (road sections or sub-areas) upon which it is to be applied, and policy-specific parameters. Alternatively, an automatic selection procedure is able to propose the objects, which fulfil a set of user-defined criteria. Safeguards have been incorporated into the policy definition module to ensure that user responses are valid. This form of user-interface is very flexible and is easily mastered by users even if they have little previous experience.

Scenario analysis corresponds to the analysis and presentation of computational results through appropriate thematic maps, diagrams and reports. The user can estimate the impacts of the set of policy measures included in the current scenario either by examining a summary report, or via an interactive report which utilizes the capabilities to produce diagrams, tables and thematic maps.

Finally, scenario comparison permits the evaluation and comparison of alternative scenarios. The evaluation of each scenario is based on a number of traffic, environmental and energy indicators. A multi-criteria analysis, where each decision is based upon judging appropriately weighted criteria, is adapted for comparing user-selected scenarios.

A Decision Support System integrated in a geographical information system (GIS) for the analysis and evaluation of different transport policies is presented by the authors from Athens (Arampatzis et al., 2004). The objective of the tool is to assist transport administrators to enhance the efficiency of the transportation supply while improving environmental and energy indicators. The Decision Support System works on three levels. The first performs the transport network analysis, the second assesses the energy consumption and pollutant emissions and the third evaluates the several policies selected. Road traffic is simulated using a deterministic, multi-modal traffic assignment model with capacity constraints. The model allows the estimation of traffic flow patterns within each link of the road network starting from the knowledge of the network characteristics and traffic demand. Energy consumption and pollutant emission calculations are based on the methodology developed by the CORINAIR working group. The evaluation of each policy scenario is based on a number of traffic, environmental and energy indicators. A multi-criteria analysis, where decisions are based upon judging according to appropriately weighted criteria, is adopted. Models are integrated in a GIS environment, which serves as the repository of the data as well as the user interface of the tool. The use of the tool is demonstrated through characteristic case studies on the Greater Athens Area in Greece. Two policy measures, one concerning the extension of the region where half of all private cars are prohibited from entering the Municipality of Athens, and the other concerning the reduction of parking spaces in the same region by 50 per cent are evaluated.

References

Arampatzis, G., Kiranoudis, C., T., Scaloubacas, P. and Assimacopoulos, D. (2004). A GIS-based decision support system for planning urban transportation policines. *European Journal of Operational Research* vol. 152, iss. 2, pp. 465-475.

Carver, S.J. (1991). Integrating multi-criteria evaluation with geographical information systems. *International Journal of Geographic Information Systems* 5(3), pp. 321-339.

Cochrane, J. and Zeleny, M. (1973). *Multiple Criteria Decision Making*, Columbia, University of South Carolina Press.

Coleman, J.S. (1971). *Foundation for a theory of collective decision.* Social Choice, ed. Lieberman, Gordon and Breach, New York.

Dasgupta, A.K. and Pearce, D.W. (1972). *Cost Benefit Analysis: Theory and practice*, London, MacMillan.

Dyer, J.S. (1972). Interactive goal programming. *Management Science* 19(1), pp. 62-70.

Fiering, B.R. (1986). *Linear programming: An introduction. Quantitative Applications in the Social Sciences*, 60, London, Sage Publications.

Fishburn, P.C. (1974). Lexicographic orders, utilities and decision rules: A survey. *Management Science* 20(11), pp. 1442-1471.

Ignizio, J.P. (1985). *Introduction to linear goal programming*. Quantitative Applications in the Social Science 56, London, Sage Publications.

Keeny, R.L. and Raiffa, H. (1976). *Decisions with Multiple Objectives: Preferences and Value Trade Offs*, New Your, John Wiley and Sons.

Kliučininkas, L. (1994). *Geographic information systems: basics and environmental applications*. University of Latvia, p. 81.

Lee, S.M. (1972). *Goal Programming for Decision Analysis*, Philadelphia, Atterbach.

Lukosius, S. and Kliucininkas, L. (2003). *Decision Support Techniques in Urban Planning. Building and Re-building Sustainable Communities*. Baltic University Press, Uppsala, vol. 2, pp. 79-84.

Moore, P.E.E. (1975). *Public Decision making and Resource Management*, Discussion Paper No.17, Toronto, University of Toronto, Department of Geography.

Nijkamp, P. (1979). A theory of displaced ideals: an analysis of interdependent decisions via nonlinear multi-objective optimization. *Environment and Planning* 11, pp. 1165-1178.

Voogd, H. (1983). *Multicriteria Evaluation for Urban and Regional Planning*, London, Pion Ltd.

Zeleny, M. (1974). *Linear Multi-objective Programming*. Berlin, Springer.

5.4 Case study VI. Decision Support Techniques in Urban Planning

Using a Geographic Information System for selecting the site of a new culture and sport hall in Kaunas, Lithuania

Source: *Lukosius, S. and Kliucininkas, L. (2003). Decision Support Techniques in Urban Planning. Building and Re-building Sustainable Communities. Baltic University Press, Uppsala, vol. 2, pp. 79-84.*

This case study will give an example of how Geographic Information Systems have been applied as a decision support tool in the evaluation of spatially distributed alternatives. Because of the difficulty in displaying GIS results, illustrations used in this report are schematic. However, real GIS analysis was used for determining distances on the specific streets.

Introduction

Kaunas is the second biggest city in Lithuania with a population of 379,000 and an area of 157 km^2. Several factors make urban planning in Kaunas problematic. First, the two rivers that flow through the city can only be crossed on a few bridges; the last one was opened in late autumn 2002 over the Nemunas river. Secondly, Kaunas is a centre with two intersecting transport axes: one is North-South the Via Baltica; the other is East-West, connecting Vilnius, the capital, and Klaipeda on the coast. Both experience intense traffic. Finally, no substantial changes are permitted either within the historic centre, or in the new centre, with its 1.6 km-long Laisves Aleja, a main avenue for both inhabitants and tourists, with offices, shops, restaurants and banks.

Basketball has always been very important and a great source of pride in Lithuania and Kaunas. During the first modern period of independence (1918-1940), Lithuanian teams were leading the European championships. Since its new independence, Lithuania regained this role quickly. The Kaunas team "Zalgiris" has enjoyed some success, in winning the European club championship in 1998 and Euroleague championship in 1999. As a result, Kaunas City Council decided to build a new Culture and Sport Hall. A new sport hall was thus projected and the urban planners were given the task of proposing the best location. It was projected to occupy a site of 25,300 m^2 and have a capacity of 6,000 seats. The new hall would also be able to accommodate other events such as concerts.

Description of sites

Initially six sites were selected for the Culture and Sport Hall (see Fig. 1). After comprehensive assessments, four lots were considered inappropriate for further analysis. Alternative III was privately owned land, and the plans of the owners did not correspond with the construction of the new Culture and Sport Hall. Alternative IV required the demolition of more than 20 private lots, which was considered too many. The traffic permeability in the neighbouring streets was considered too low at alternatives V and VI.

Fig. 1: Six alternative sites for the location of the new Culture and Sport Hall in Kaunas

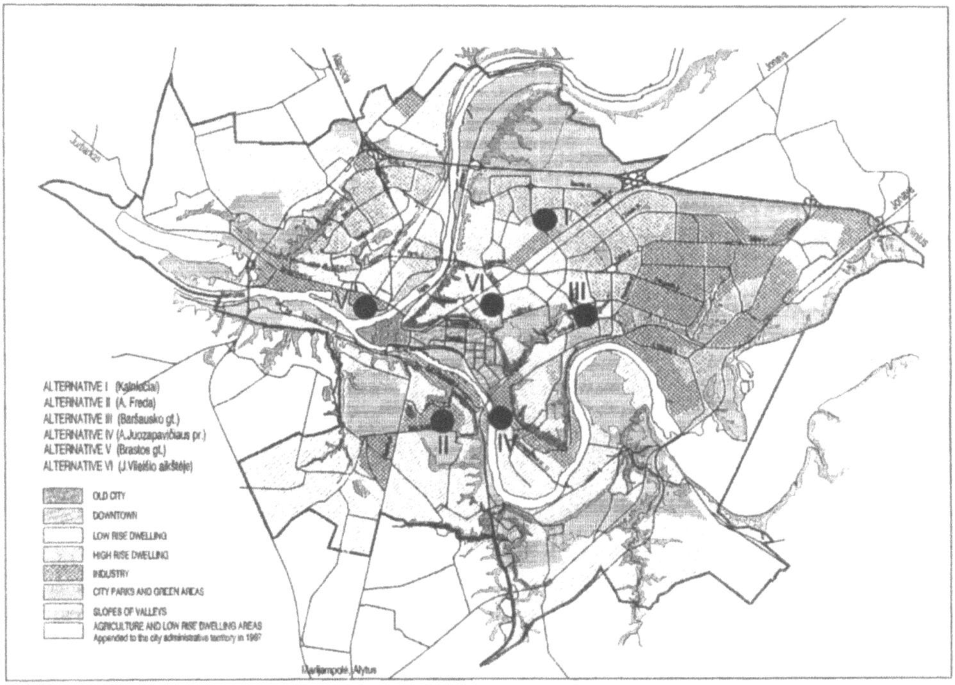

More detailed analyses were therefore undertaken for alternatives I and II. Alternative I was situated in one of larger parks in the city, Kalnieciu Parkas. Alternative II was located at the Aukstoji Freda, with good traffic connections, especially considering the plans for a new bridge to be built over the Nemunas river.

The two sites were compared and evaluated according to several different sets of criteria. These included accessibility (e.g. distance to city centre), status (e.g. technical infrastructure), communication possibilities, and functional and architectural conformity.

Criteria for evaluation of alternative sites for the Culture and Sport Hall

A. Availability
A1. Distance to city centre
A2. Distance to transport terminals
A3. Distance to highways
A4. Distance to residential areas

B. Present status
B1. Size, elevation, greenery
B2. Legal (property ownership)
B3. Character of the buildings
B4. Available engineering networks

C. Communication possibilities
C1. Network of neighbouring streets
C2. Pedestrian walkways
C3. Public transport
C4. Individual and group transport

D. Functional conformity

E. Architectural conformity

Alternative I. Kalnieciu Parkas

The site is situated in the north-eastern part of Kaunas, in Kalnieciu City Park. The development of the City Park began in 1975; today the park has an area of 21.8 ha. The site is surrounded by the Kalnieciu (30,000 inhabitants) and Eiguliu (25,000 inhabitants) residential districts.

The site is 1.5 km away from centre of gravity of the city (calculated assuming that the surface of the city has equal density, also called the "weight" centre (see Fig. 2). The distance from the city centre (downtown) is about 4 km (see Fig. 2). The network of regular buses, trolleybuses and microbuses is well developed.

The sites were also examined with respect to the main work-to-home traffic corridors, as well as the main suburbs-to-city traffic corridors. Thus the location, with respect to the existing and prospective "inter-city" bus stop, is given in Fig. 3.

The train station and bus terminal are situated 4.5 km from the site, and the distance from Kaunas (Karmelava) airport is 6.2 km.

Analysis also involves the consideration of the mean prospective number of passengers using public transport during peak hours. On one side the site borders Savanoriu Avenue, one of the main transport routes in the city, connecting the city centre to the Vilnius-Klaipeda highway. On the other side of the site, the very busy P. Luksio Street and S. Zukausko Street are located. The distance to the nearest suburban bus stop is 600 metres. Distances to the bus stops are shown in Fig. 4.

Alternative II. J. Bakanausko Street

This site is situated in the south-western part of the city on the left bank of the Nemunas. After the construction of a new bridge across the river, the street was widened to six lanes of traffic with a possible traffic intensity of 3,000-3,5000 vehicles per hour. The street has become the main traffic artery connecting the south-western and north-eastern parts of the city. Neighbouring the site are the Botanical Gardens, industrial sites, storage facilities and low rise dwellings.

The site is located 4.5 km from the city "centre of gravity" (see Fig. 2). The distance from the city centre (downtown) is about 2.5 km. The distance from the railway station and bus terminal is 3 km. Kaunas (Karmelava) airport is 14 km from the site. In the frame of development of the Master Plan of Kaunas, the prospective calculated transport flows in both directions during peak hours were analysed.

Fig. 2: Location of two alternative sites in relation to the city "weight" centre

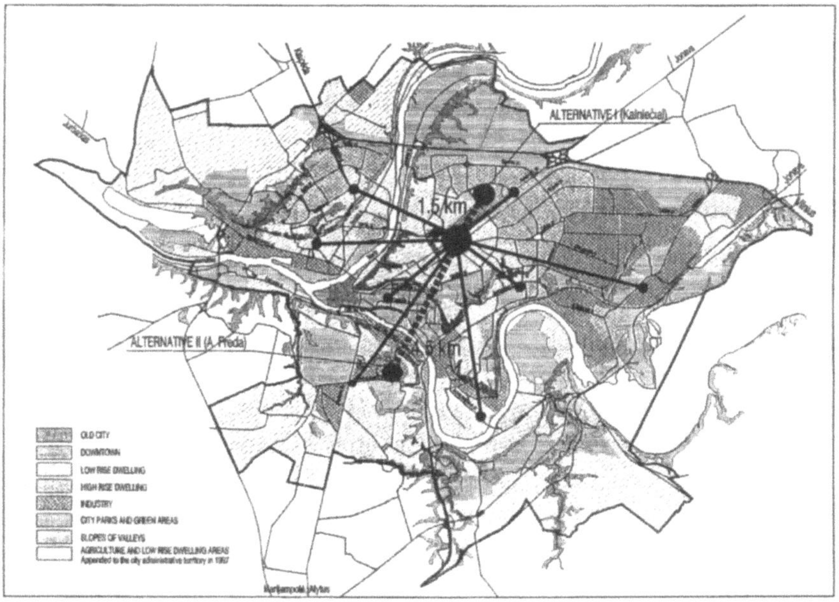

Fig. 3: Location of two alternative sites with respect to the existing and prospective "inter-city" bus stops

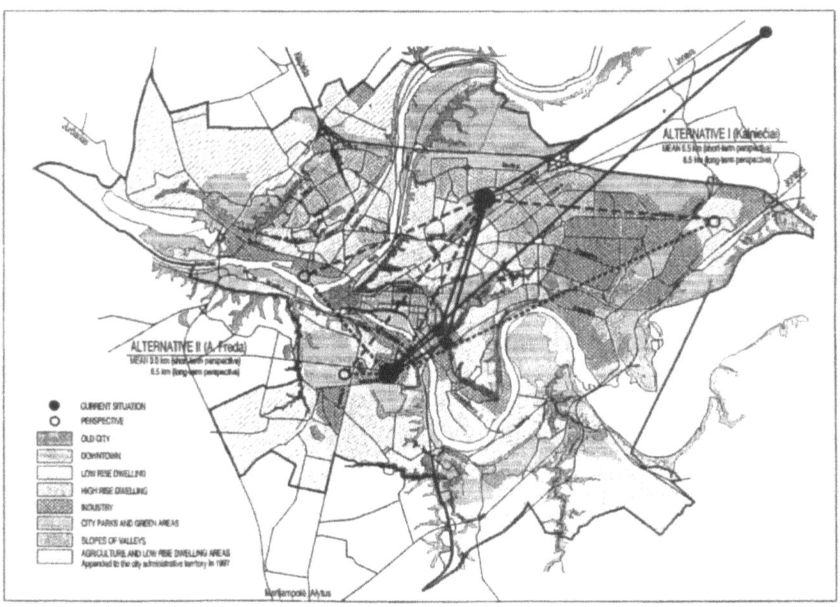

Fig. 4: Distances from the two alternative sites to the suburban bus stops

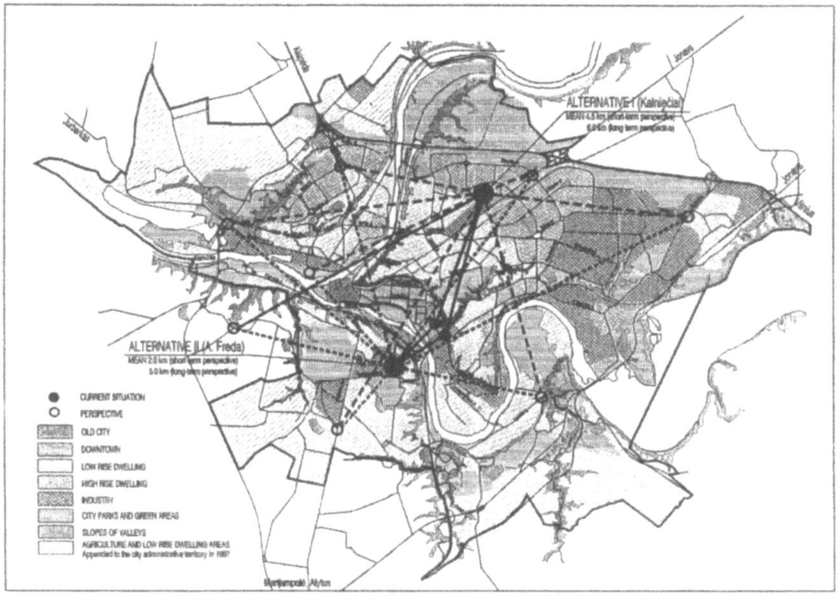

Applying decision support techniques – multi-criteria evaluation of convenient access

To make the decision-making process participatory and effective, decisions require a technique that can directly incorporate the very special knowledge that stakeholders possess (Dennsham and Goodchild, 1989; Smit and Smit 2003; Walker et al., 2003). One such a technique is multi-criteria evaluation.

This technique was applied for the following four factors.

1. *Proximity to the bus terminal/railway station.* The bus terminal and railway station in Kaunas are situated near each other. It is assumed that a considerable number of visitors, non-residents of Kaunas, will use inter-city buses or trains.
2. *Proximity to the city centre.* Many visitors will travel to the site from the city centre, since there is a high concentration of supporting services and governmental offices. A heavy traffic flow towards the city centre is expected after events at the Culture and Sport Hall.
3. *Proximity to "weight" centre.* Closeness to the city "weight" centre will be advantageous for remote citizens. Travel from the same concentric circle will require quite equal time expenditure; visitors will therefore have the same chance of reaching the site at the same time.

4. *Proximity to the suburban bus stops.* An analysis of the perspective concept for the development of suburban bus stops gives additional information on the short and long-term allocation of bus stops.

The next step was to create a pairwise comparison matrix in which each factor is evaluated for its importance relative to every other factor. This can be done by a single individual or by a group of decision-makers. In our case, the working group consisted of experts in the fields of environment, energy production, urban planning and healthcare. The group has represented a broad spectrum of interests related to the new Culture and Sport Hall. A nine-point rating scale was used (Table 1). A symmetrical matrix incorporating four factors was set up (Table 2).

Table 1: A nine-point rating scale

1/9	1/7	1/5	1/3	1	3	5	7	9
extremely	very strongly	strongly	moderately	equally	moderately	strongly	very strongly	extremely

less important *more important*

Table 2: Matrix for comparing factors in multi-criteria evaluation

	Station	City centre	City weight	Bus stops
Station	1			
City centre	1	1		
City weight	5	3	1	
Bus stops	1/3	1/5	1/7	1

Since each factor is of equal importance to itself, the diagonal (Table 2) is filled out with the figure 1. The pairwise comparison then begins from the second row in the first column. One individual from the group judges the importance of the row factor relative to column factor by working with a question such as: "Relative to the station factor, how would you judge the importance of the centre factor?". The comparison is given maximal rating of five. There follows a discussion, after which the group of decision-makers comes to a consensus. The procedure continues. Table 2 shows the result of one such exercise. Once the matrix is completed, a pairwise comparison is used to estimate factor weights. The results were as follows:

1. Railway/Bus station factor 0.1582
2. City "downtown" factor 0.2047
3. City "weight centre" factor 0.5803
4. Bus stop factor 0.0568

To complete the estimation of alternatives, we need to apply rank ordering of distances, calculated during previous GIS operations. Now we are able to estimate alternative sites from the point of view of convenient transport access. The scores for the two alternative locations are 7.42 for the park site and 6.89 for the site at J. Aukštoji Freda. Alternative I, Kalnieciu City Park, is thus more attractive in terms of reaching the new Culture and Sport Hall using different means of transport. However, since the difference between the total weights of alternatives was negligible, it was considered that the Kalnieciu City Park site was not significantly better in comparison with the Aukštoji Freda site.

Table 3: Ordering of distances, calculated by GIS for each site used in multi-criteria evaluation

Distance in km	9-10	8-9	7-8	6-7	5-6	4-5	3-4	2-3	1-2	0-1
Score	1	2	3	4	5	6	7	8	9	10

Table 4: Estimation of alternative sites using multi-criteria evaluation of four factors of convenient access to transport

	Factor weight F_w	**Score** S_1	**Alternative I** $F_w \times S_1$	**Score** S_2	**Alternative II** $F_w \times S_2$
Station/terminal	0.15	4*	0.63	8*	1.26
Centre	0.20	6	1.22	8	1.63
Weight centre	0.58	9	5.22	6	3.48
Suburban bus stops	0.05	6*	0.34	9*	0.51
TOTAL			7.42		6.89

* *A long-term perspective was not taken into consideration*

Calculating pedestrian access and requesting parking space Two more analyses were done to judge the accessibility of the projected Culture and Sport Hall: the expected number of visitors by car and bus vs. the number of expected visitors on foot.

These two figures were considered together as they are interdependent. Both were calculated on the basis of the total number of visitors. It is assumed that the population from which visitors to the Culture and Sport Hall will be recruited is around 600,000. Of this figure, 400,000 is the population of Kaunas. Another 200,000 in the larger region will be potential visitors at sport and culture events. As access to automobiles in the future is expected to increase from the current level by around 180 cars to 500 vehicles/1000 inhabitants, many of the visitors will thus come in private cars, and parking spaces are required.

But let us first look at the number of potential visitors on foot. The number of inhabitants living within walking distance of the new Culture and Sport Hall was a crucial factor.

The proposed park site, Alternative I, is surrounded by residential districts of Kalnieciu (30,000 inhabitants) and Eiguliu (25,000 inhabitants). Approximately 15,000 inhabitants live near the Savanoriu Avenue traffic route. In total 60,000 inhabitants live within walking distance of the proposed Culture and Sport Hall; that is, within a 15-20-minute walking distance, or 1.5-2.0 km. Location with respect to the pedestrian routes is presented in Figure 5 (figures indicate the comparative number of pedestrians). This corresponds to 10 per cent of the total population expected to use the hall. At a single event, up to around 600 visitors (10 per cent of 6,000) are thus expected to belong in this category.

At the alternative site at Aukštoji Freda, there are only 18,000 inhabitants living within 1.5-2.0 km distance. 18,000 constitutes 3 per cent of the population basis, and thus an expected 200 visitors.

Fig. 5: Locations with respect to the main pedestrian routes

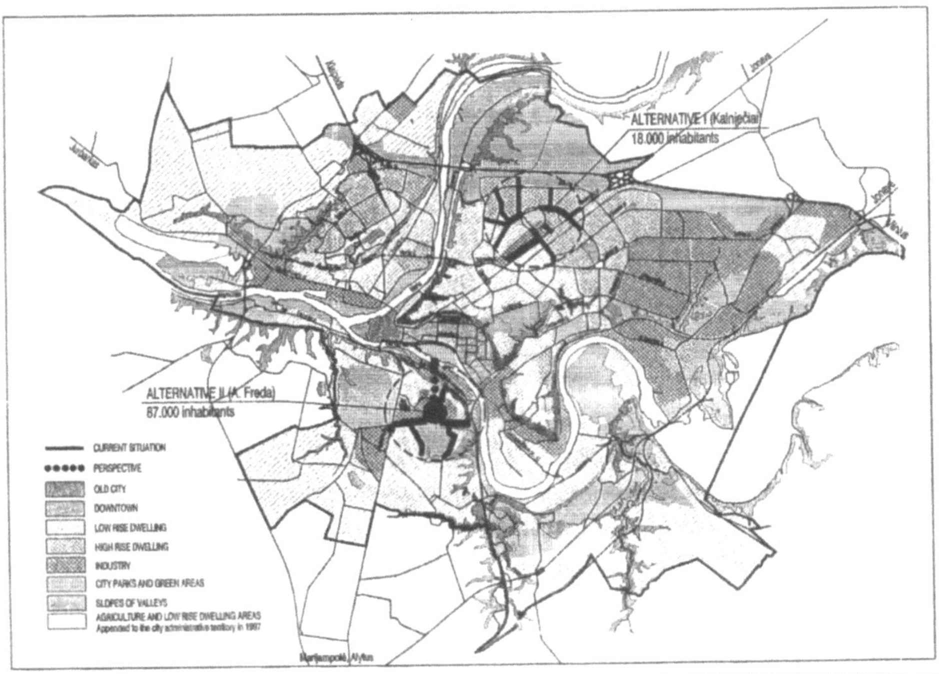

Based on studies of visitors to sports events, planners assume that 50 per cent of visitors (up to 3,000) will use private cars. On average, every individual vehicle is assumed to bring 3 passengers. Thus the maximum expected number of cars is 1,000. Likewise the number of visitors using city buses was predicted to be 20 per cent, or 1,200 visitors. Passengers travelling by intercity buses will comprise 15 per cent, i.e.

900 visitors. If each intercity bus takes 30 passengers, 40 parking spaces will be required for intercity buses. The total number of parking spaces projected was thus 1000 + 50 = 1,050 light duty vehicles and 40 + 40 = 80 heavy duty vehicles, where the additional 50 cars and 40 buses were reserved places for service vehicles.

At the Aukštoji Freda site, an additional 800 people were assumed to come by car. Thus, assuming a constant relationship between buses and private cars, an additional 270 parking spaces for cars and 12 for buses would be needed.

Discussion

Optimising access to the hall in relation to transport, including links to the train and bus station, "downtown" and city weight centre, as well as bus stops, clearly points to the Kalnieciu Park location, Alternative I, as the best.

Optimising pedestrian access and thereby reducing both needs for parking spaceand motor transport also strongly favours the Kalnieciu Park location. The lowest acceptable number of pedestrian visitors was set to 5 per cent, i.e. 300 visitors, which is only possible with the park location.

It seems that the arguments suggest that it would not be a good decision to place the hall in the less developed area at the Aukštoji Freda site, Alternative II. Alternatively, if the hall is built at the Kalnieciu City Park, some measures need to be taken to protect the green character of the city.

However, the urban planners did not need to fight with this dilemma. Due to limited financial resources, the municipality of Kaunas abandoned the construction plans.

In 2008, construction began on the new Culture and Sport Hall, in a new location that was not analysed in this case study.

References

Dennsham, P.J. and Goodchild, M.F. (1989). Spatial decision support system: A research agenda. In *GIS/LIS'89*. Orlando, vol. 2 American Congress on Surveing and Mapping, Bethesda, pp. 707-713.

Smit, W. and Smit, I. (2003). How much support can we expect from decision support systems? *International Journal of Technology, Policy and Management* 3 (3/4), 251-261.

Walker, W.E., Harremoes, P., Rotmans, J., van der Sluijs, J.P., van Asselt, M.B. A., Jamssen, P., Krayer von Krauss, M.P. (2003). defining uncertainty: a conceptual basis for uncertainty management in model-based decision support. *Integrated Assessment* 4, 5-18.

5.5 Life Cycle Assessment

5.5.1 Life Cycle Impact Assessment

Life cycle assessment (LCA), is a fairly new technique for evaluating the environmental effects of a product or a service over the entire period of its life cycle ("from cradle to grave"). Thus, when performing an LCA, all emissions and resource consumption which enter or leave a life cycle are translated into the environmental problems that they may potentially contribute to. The two terms "environmental effects" and "life cycle" both need to be properly understood.

Environmental effects are the consequences of physical interaction between a system studied and the environment. In practical use, all environmental effects are represented by several categories of environmental problems. The most commonly used are:

- Resource depletion
- Global warming
- Ozone depletion
- Human toxicity
- Ecotoxicity
- Photochemical oxidation
- Acidification
- Eutrophication
- Land use
- Others (including solid waste, heavy metals, carcinogens, radiation, species, extinction, noise).

There are many shortcomings in the applications of LCA techniques. Up to the present time, the main focus of research in LCA has been on developing LCA as a tool rather than a system (Hauschild and Wenzel, 1998). Hence, there has been an emphasis upon assessment of potential environmental effects. Relatively little attention has been paid to, for example, the process of defining alternatives for consideration in an LCA, or choice of weighting factors. Further research could allow us to develop LCA as a system capable of assessing the environmental impact per function, system needed or money spent. Such a new approach would allow us to search for conditions for reaching maximum cost-effectiveness, with respect to the environment, for a function or product. Then the ratio of system cost/environmental impact could be maximised (Stavenuiter, 2002).

Today's LCA approaches are valid only for incremental changes in the product of interest and defined geopolitical regions. To use LCA to provide answer to long-term planning issues is more difficult. Radical changes in technologies, legislative constraints or policy goals need to be anticipated. Future developments of key technologies and entire economic sectors will thus affect the outcomes of the LCA (Frisknecht, 1997).

Current LCA integrates over time and space. A desegregation of these two parameters is needed to ensure more precise results. Environmental impacts depend on location; for example acidification is very different in different places. Desegregation of the inventory of impact as to location is a matter of practicability. It has been done in several studies in order to allow for a differentiated impact assessment. Desegregation in time is needed to allow for a differentiated impact assessment. In current practices LCA is supposed to support decision-making and affect future decisions, whilst, for an actual system, a substantial part of the processes has already taken place.

Qualitative (approximate) *LCA methods* do not use systematic computational procedures to assess the environmental profile of the system under study. They analyse the life cycle in environmental terms directly on the basis of emissions released and the consumption of raw materials. Assessing the seriousness of the impact requires training and extensive knowledge. A decisive role is played by the relevant experiences of the expert carrying out the evaluation. There are a number of different *quantitative LCA techniques*. These are in practice applied as a group of methods which use classification, characterisation, normalisation and weighting. The methodological framework of all the LCA techniques is based on ISO standards 14040-43.

5.5.2 Quantitative LCA

The first step in an LCA is a raw assessment to determine critical points in the life cycle and find directions for further studies. Such a quick analysis is called *screening*. *Goal definition and scope* is crucial for all the other phases. These include gathering data, which involves building a model of the life cycle and choosing appropriate environmental effects to consider local and global impacts. The next step is *inventory analysis and allocation*. The data from an inventory table has to be processed to attain a higher level of *aggregation*. Ideally the aggregation process results in a meaningful single score. To achieve this, the ISO standards advise a four-step procedure:

- Compulsory steps: classification and characterisation,
- Optional steps: normalization and weighting.

The last step, the *improvement assessment phase*, is performed in accordance with the goal of the study and on the basis of results from the impact assessment phase (Zbicinski et al., 2006).

The *goal definition and scope phase* requires a clear description of the goals of the study and the development of its scope. The scope of the study sets requirements as to the desirable level of detail. The main issues to be considered at this stage are:

- Purpose of the study: why is the analysis being performed? What is the end use of the LCA? To whom are the results addressed?
- Specify the product (service) functional unit to be investigated (functional unit),
- Scope of the study: depth and breadth (system boundaries).

A *functional unit* provides a reasonable point of reference when comparing different products. It should be clearly defined.

System boundaries are the next vital tasks to be defined in the goal and scope definition step. The necessity of defining system boundaries results from the fact that the main technique applied in any LCA is modelling. Models of environmental mechanisms are created to translate inflows and outflows from the life cycle into the environmental impacts they may contribute. For example, SO_2 emissions could increase acidity. This, in turn, can cause soil and water damage, influence the quality of the ecosystem and have harmful effects on living conditions for animals and plants.

Inventory analysis and allocation is the core of an LCA and a common feature of any LCA. During this phase all the material flows and energy flows, and all the waste streams released into the environment over the whole life cycle of the system under study, are identified and qualified. The final result of the inventory analysis is an inventory table. A typical LCA inventory table consists of a few hundred or more items. They might be grouped into categories: raw materials, emissions to air, water, soil, solid emissions, non-material emissions (noise, radiation, land use), etc.

In the *allocation* step, emissions and material consumption is divided between several products or processes. Very often, a process fulfils two or more functions or gives two or several usable outputs. They are multi-output processes. Then we have to determine which part of the total emissions and material consumption should be attributed to each specific product. Petrol production can serve as an example of a multi-output process. It provides several products in fractional distillation of crude oil: not only petrol but also kerosene, diesel oil, and heavy fuel oil. The same applies to multi-input processes.

Classification of impacts is the first step to higher aggregation of the data. Inflows and outflows from the life cycle are gathered in a number of groups representing the chosen impact categories. The inventory table is rearranged in such a way that the amount of relevant emissions or the material consumption is listed under each impact category.

Characterisation provides quantification of different substances and their contribution to the different impact categories. During the characterisation step the relative strength of the emission is evaluated and contributions to each environmental problem are quantified. This is done by applying so-called *equivalence factors*. An equivalence factor indicates how much more a given compound contributes to a problem in comparison to a chosen reference substance.

The final result of the characterisation step is a list of potential environmental impacts. This list of effect scores, one for each category, is called the *environmental profile* of the product or service.

Normalisation is a procedure to allow us to compare impact categories among themselves. The results from the characterisation step cannot be compared since they are usually presented in different units (e.g. CO_2 eq., SO_2 eq.. Normalisation is performed to make the effect scores of the environmental profile comparable. The normalised effect score is the percentage of a given product's annual contribution to that effect in a certain area.

Weighting is the step in which the different impacts categories are weighted so that they can be compared among themselves; i.e. the relative importance of the effects is assessed. The weighting (valuation) is performed to obtain a single score representing the environmental impact of a product or service. This is still a subjective process, although much effort has been spent in recent years to work out a scientific basis for weighting. In practice, weighting is performed by multiplying a normalised environmental profile by a set of weighting factors, which reflect the seriousness of a given effect. If each impact category is provided with a factor according to its environmental significance, an environmental profile can be expressed in a single *environmental index*. An environmental index is a sum of the numbers which a weighted environmental profile consists of. Once the environmental indices are calculated, comparisons of products (services) are easy.

Improvement assessment thoroughly analyses all the results obtained during previous methodological stages and gives answers to questions asked in the goal and scope definition phase. In this phase a detailed report is also drawn up to present the results in the most informative way possible, suited to the intended audience. There are several issues to cover.

Uncertainty. The LCA methodology still has many weak points and strongly depends on the quality of data, which frequently is extremely hard to obtain. All these factors must be highlighted when drawing conclusions from the analysis, in order not to mislead the audience. There are two main sources of uncertainty. First is the quality of the data: data often comes from different sources, estimates, assumptions, theoretical calculations, etc. Secondly, any LCA includes subjective choices which cannot be avoided: uncertainty is part of the model, for example in terms of system boundaries, allocation rules, or characterisation models.

A *sensitivity analysis* is carried out to check how stable the results are. It should be proved that input data and methodological choices do not influence the results of an LCA too much. During sensitivity analysis the LCA is recalculated and the outcomes are compared. In this way the critical points in an LCA are identified.

References

Friscknecht (1997). *Goal and Scope Definition and Inventory Analysis*. LCA-NET Theme report, ETH: Zürich, Switzerland.

Hauschild, M., Z. and Wenzel, H. (1998). *Environmental assessment of products*, vol. 2: *Scientific background*. Chapman and Hall: London, UK.

Stavenuiter, J. (2002). *Cost Effective Management Control of Capital Assets*. Asset Management Control Research Foundation: Medemblik, the Netherlands. (http:www.amc-rf.com).

Zbicinski, I., Stavenuiter, J., Kozlowska, B. and van de Coevering, H.P.M. (2007). *Product design and Life Cycle Assessment*. Uppsala, The Baltic University Press, p. 308.

5.6 Integrated assessment of environmental impacts

5.6.1 Transport planning and Strategic Environmental Assessment

An indicator can be defined as the basic instrument to summarise the information necessary to evaluate a system. Thus, the research on indicators comes from the need to translate in a simple and immediate way, through a number, a word or a graphic sign, complex phenomena characterized by different types of data in a continuous process of aggregation and disaggregation of the information.

It can be argued that indicators and indexes (aggregation of indicators) represent an empirical model of the reality, which assumes that a complex phenomenon can be depicted by the limited set of variables.

The selection of indicators is a very difficult task, due to the many indicators existing in literature. Since the choice of indicators must be supported by a scientific and transparent approach, this chapter will provide an outline for the selection of indicators which are relevant to the Strategic Environmental Assessment (SEA) of transport plans and programmes. The selection of indicators is the most delicate part of assessment, because the ability to consider interactions between environmental (air pollution, noise, land use, etc.), technological and policy (pricing, taxation, etc.) aspects is a guarantee for successful traffic planning. The difficulty of achieving a reliable assessment of the environmental impacts is found in joining transport and environmental models in different levels of detail.

The different typology of pollution and impacts, and the complexity of environmental impact mechanisms, are so high that the decision-makers need support in making choices based on a small number of simple, but strict and scientific, elements. *Mobility*, *land use* and *environment* are the key groups of data from which indicators relevant to the transport planning could be derived.

To define which transport indicators are suitable for evaluating the effects of the transport plans and programmes, it is fundamental to individuate some of the main criteria helping decision-makers to choose the most appropriate indicators. The indicators proposed should be easily computable and should allow the definition of trends due to the transport planning and policies adopted. In this way it is possible to compare indicators calculated each year and monitor the effects of the plan. On the macro level, COST 350 action has proposed the following indicators:

- Accessibility;
- Level of Service (LOS);

- Rate (%) cycle paths surface/roads surface;
- Rate (%) reserved roadway surface/plan area surface;
- Rate (%) surface for public transport/reserved roadway surface;
- Network extension (km) of public transport lines;
- Rate (%) number of interchange parking spaces;
- Length of roads (km) inside ecologically rich areas (National parks, Natura 2000, etc.);
- Change of the land use value;
- Proximity (m) of transport infrastructure to areas of ecological, cultural, or social significance, or areas of natural beauty;
- Percentage of people living in the areas alongside the infrastructure (250 m each side) with respect to the population living in the plan area;
- Rate of area dedicated to interventions focused to reduce transport demand (e.g. pricing policies, zone 30)/plan area surface.

The *DPSIR-scheme* provides a conceptual framework for analysing the interrelated factors that have an impact on the environment and is useful for any environmental assessment. This approach proceeds to a more micro approach going into great detail. The work of the European Environmental Agency (EEA) is built around a conceptual framework known as the DPSIR assessment framework (TERM 2005): D – Driving Force, P – Pressure, S – State, I – Impact, R – Response.

Fig. 12: DPSIR in the transport sector

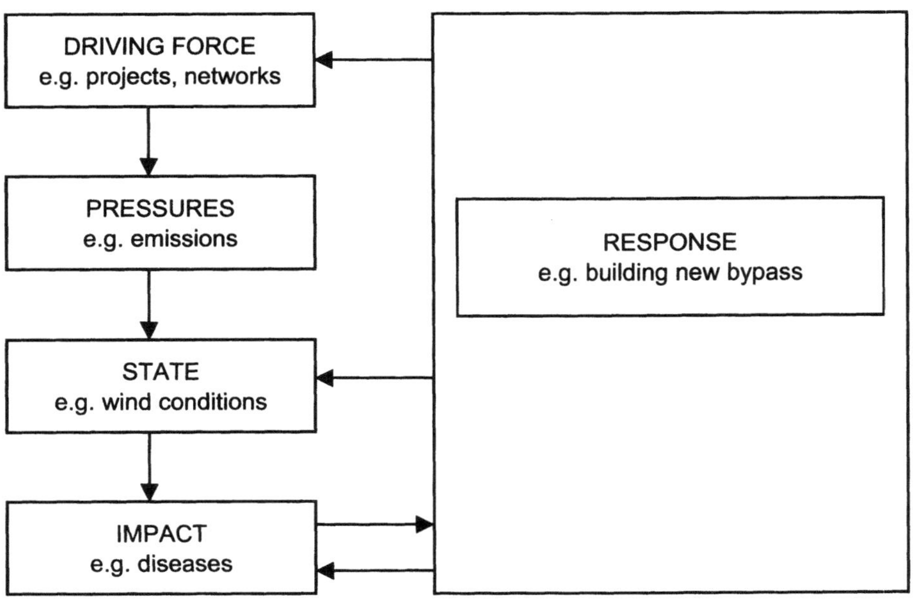

Figure 12 clarifies the application of this approach in transport planning. Driving force means the entire traffic network of the region being examined. This force causes effects (pressures, for example emissions), which have consequences depending on the relevant environmental situation (state, for example wind conditions). When human health and the environment are exposed to emissions, the results are shortened life expectancy and the degradation of ecosystems. Part of the answer (for society, for the administrative authorities, for policy-makers) to these changes in the environment (response) with respect to the whole transport plan/programme could be, for example, the building of a new bypass. The observed effect on the environment can be also positive: for example noise reduction in town as a result of the new bypass.

The use of the DSR framework does not mean that it is possible at this stage to identify any causal relationships between driving force, state, and response indicators. Rather it should be seen a way of categorising indicators to fit the needs of planners. Nevertheless, as time series data are collected and analysed for these indicators, patterns and relationships may emerge that highlight connections and interactions among indicators. The search for a set of indicators is a process involving trial and error. It will also eventually involve moving from linear to dynamic models of transport planning.

Based on findings from COST 350 (Calderon et al., 2009) action indicator selection criteria have been classified in two groups: general criteria and specific criteria. Five general criteria have been proposed:

- significance,
- completeness,
- simplicity and applicability;
- scientific validity;
- transferability.

In addition to these general criteria, specific criteria may be added:

- European-rules-oriented approach,
- transport-oriented approach,
- SEA-oriented approach,
- Decision-making-oriented approach.

The scheme has allowed the defining of ten criteria used to select the indicators. Each criterion is evaluated according to different questions and a mark is given for the different questions. The magnitude of the response for all the questions is defined as: 1 (negative answer), 2 (positive answer), 3 (very positive answer). The aggregation of the different marks, where more questions are considered within a criterion, is based on a simple addition of the different results, giving a final score. It was stated that an indicator should satisfy a minimum of 50 per cent of each criterion (that means 50 per cent of the questions relative to each criterion) to be adopted. If one indicator does not satisfy this minimum of 50 per cent for one criterion, the indicator cannot be selected. Authors have considered seventeen impacts, where every impact corresponds to an indicator which is then divided into three corresponding levels of information: low, intermediate and high data availability. Six indicators were defined as the most relevant ones to be taken into consideration (see Table 14). For reasons of simplicity only indicators with a high level of information availability are presented. The key environmental impact indicators for a strategic valuation are listed in Table 14.

Table 14: The key environmental impact indicators proposed by COST 350

No	Impact	Indicator at the highest information level
1	Land uptake	Natural habitat area lost Domestic and recreation area lost Sealed area
2	Fragmentation of habitats	Endangering of populations of (representative) target species
3	Disturbance from noise	Number of people affected by excessive noise levelsor proximity of sensitive habitats
4	Climate change	CO2 emission
5	Toxicity	Number of people or size of protected areas exposed to excessive toxic or ecotoxic emissions of heavy metals, POPs, PM, NOx, SOx
6	Accidents	Number of killed, seriously or slightly injured persons due to accidents

The key impact indicators for road transport assessment recommended by European Environmental Agency (EEA) are presented in Table 15. The list of indicators was developed in the framework of the Transport and Environment Reporting Mechanism (TERM, 2005).

Table 15: The key environmental impact indicators proposed by EEA

No	Impact	Indicator
1	Climate change	Emissions, greenhouse gases
2	Air pollution	Potential concentrations of particulate matter
3	Noise	Annoyance
4	Accidents	Traffic fatalities
5	Protection of biodiversity (terrestrial)	Unfragmented areas
6	Energy resources	Final energy consumption
7	Land as resource	Land-take by transport infrastructure

The main difference between the COTS 350 model and the TERM model is not in the type of impacts proposed (they are actually very similar), but with regard to relative indicators corresponding with every impact.

5.6.2 Impact scoring and aggregation

Impact significance is assessed through *impact scoring*, which is the process of assigning a quantitative or qualitative value to environmental impacts arising from transport measures. The scores given reflect the significance of the impact. In good practice, a pre-defined structured framework or a list of significance criteria is used to reach transparent, acceptable and auditable results. Such frameworks or significance criteria can be prepared for most, if not all, environmental topics. Essentially, the criteria can be established using regulatory norms, distance to target and other quantitative or qualitative descriptions of the environmental impact.

At the impact assessment stage, the predicted impacts are translated into statements of significance. Prediction of the magnitude of likely future impacts and the sensitivity and importance of the receptor (the environment or humans) leads to a possible statement about the significance of the impact.

While the practice of establishing and defining significance criteria within project level EISs is often not carried out to the highest standards, the application of significance criteria to SEA is at a very early stage of development. While assigning positive or negative impact scores may appear to be a simple procedure, it can raise questions of consistency and transparency of the scoring. The question of who sets the criteria and whether there should be a standardised set of significance criteria applied nationally across all tiers in the transport planning process, in order to provide decision-makers with an equal basis on which to compare plans, is yet to be considered, let alone answered. Whereas a general framework of significance criteria should be considered necessary, this report advocates that impact scoring should always be undertaken on a case-by-case basis for individual plans and programmes.

An impact of the same magnitude is expected to give rise to different levels of impact significance, depending on the likely evolution of the environment without the implementation of the plan or programme (see Fig. 13).

Step 1: Determining the sensitivity/importance/value and the bearing capacity of the receiving environment.
A pre-defined scale (with explanation) for value and vulnerability of the area likely to be affected could help reach clear results.

Step 2: Measuring the current state of the environment and predicting the future magnitude of the impact. A pre-defined scale (with explanations) for magnitude of the effects is necessary.

Step 3: Assigning scores and estimating the overall assessment score.

As is obvious, *aggregation* is part of the assessment process whereby decision-makers opt for a preferred transport alternative. It is also evident that aggregation takes place close to the final stages of decision-making, and can thus be seen to aid those responsible for the final decision; likewise, it can be seen as a tool for democratic public participation in the process of assigning importance scores to all impacts stemming from alternatives under consideration. No. universally accepted definition for impact aggregation stands out. The ideas set out so far, however, as well as the outcome of the literary review undertaken, have resulted in the following definition being proposed.

Aggregation is defined as the combination of impact scores into a final/overall assessment score on the project and strategy/plan scale within individual topics. Those indexes may be further developed, jointly considered, and information distilled, in order to report to both the public and decision-makers.

Ideally, an unbiased decision ought to be based upon an objective index, reflecting the "value" of the alternative under consideration. However, the difficulty of reaching such an undisputed index has led to the acceptance that the purpose of aggregation may not necessarily be to produce a single overall environmental index to be compounded with a socio-economic one, but, more simply, to bring together all individual impacts across the plan area to be presented to decision-makers and stake-holders at large. Due to all these shortcomings, no formalised method for impact aggregation is available. Different countries apply ad hoc procedures which often reflect evaluators' experience rather than unchallengeable procedures (Goger, 2005).

Fig. 13: Identification of significances of impacts (Source: Glasson et al., 1999)

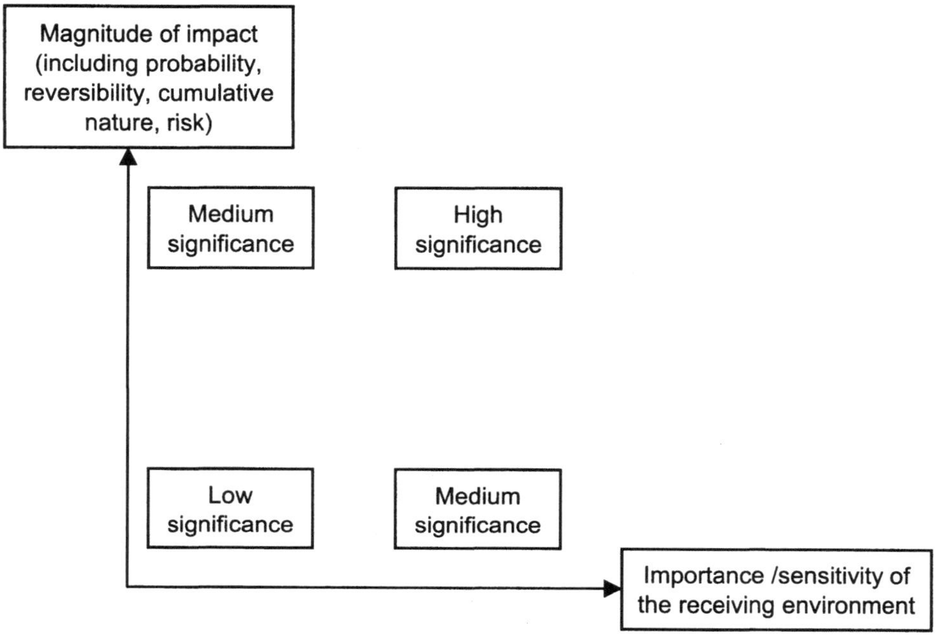

References

Calderon, E.J., Pronello, C. and Goger, T. (2009). *Integrated assessment of environmental impacts of traffic and transport infrastructure*. COST office, p. 405.

Goger, T. (2005). *Three sets of indicators for SEA of traffic and transport infrastructures*. SEA Indicators workshop 2. November (Ireland), EPA, p. 7.

TERM 2005 (2006). *Indicators tracking transport and environment in the European Union*. EEA report No 3.

5.7 Case study VII. Integrated assessment of damage from urban motorised mobility induced ambient air pollution

***Source**: Kliucininkas, L. and Macionyte, E. (2008). Integrated assessment of damage from urban motorised mobility induced ambient air pollution, in The international Conference "Environmental Engineering": Selected papers, vol. 3. Ed. by Čygas, D. and Froehner, K.D. May 22-23, 2008, Vilnius, Lithuania. Vilnius: Technika, pp. 987-992.*

This case study presents the application of the "emda vector" approach to the integrated assessment of urban transport effects on environment, taking into consideration vehicular emissions and damage factors. The exhausts of NOx, SO_x and PM_{10} released by motor vehicles were transformed into damaging effects on human health and ecosystems.

In order to derive integrated indicators for the urbanised area, the authors applied the Eco indicators 99 method and used the EMEP/CORINAIR vehicular emission factors database. Later integrated indicators were combined with calculations of urban vehicle kilometres of travel (vkt). This approach was tested in the Lithuanian city of Kaunas.

Introduction

The relationships between transport and the environment are multidimensional. An understanding of the reciprocal influence of transport-environment and the integrated assessment of environmental impact of traffic and transport infrastructure in order to assist policy makers is the primary task for the scientific community (COST 350, 2006).

European Commission guidance (Directive 2001/42/EC, 2003) gives some direction regarding how to assess the significance of environmental effects. It clearly states that the assessment of significance should take into account: i) the characteristics and value of the receiving environment, and, ii) the magnitude of the impact.

Emda vector

The authors have introduced composite entity, which comprises qualitative and quantitative characteristics of vehicular emissions. The weight **emda** (**em**issions + **da**mage) was derived in order to estimate impacts of vehicular emissions on human health and ecosystems. On the common basis the emda can be used to describe any chemical emitted by the motor vehicle. Emda is not a scalar, but a vector, which means that composite entity not only has magnitude, but also direction. The emda value is estimated as vector summation of two constituents in perpendicular axes (see Fig. 1). Each emda vector is comprised of emission and damage vectors. As a result, emda could be applied using vector algebra as well as operations of mathematical physics. In other words, the "vector approach" extends analytical possibilities, if compared to scalar values.

Fig. 1: Identifying impact significance

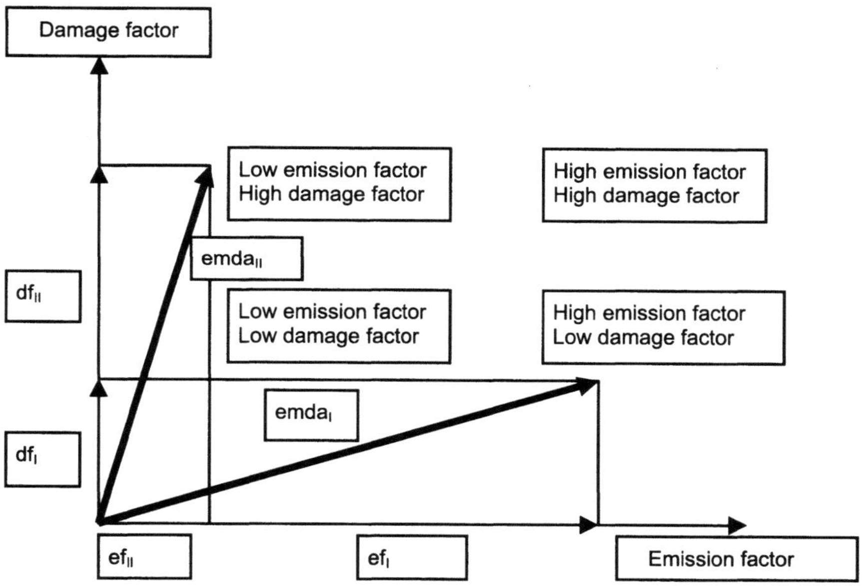

Since the authors of this article have considered only local effects of ambient air pollution, the following categories of impacts are defined: i) respiratory effects on human health (inorganic compounds), ii) acidification and eutrophication effects on ecosystems. The contaminants that have direct influence on those effects are: SOx, NOx, PM. It should be noted that the same contaminant has different influences on human health and the acidification and eutrophication of ecosystems. Thus, the two categories of impacts mentioned above should be treated as separate vector fields (see Fig. 2).

Fig. 2: Schematic view of impacts on human health and ecosystems quality

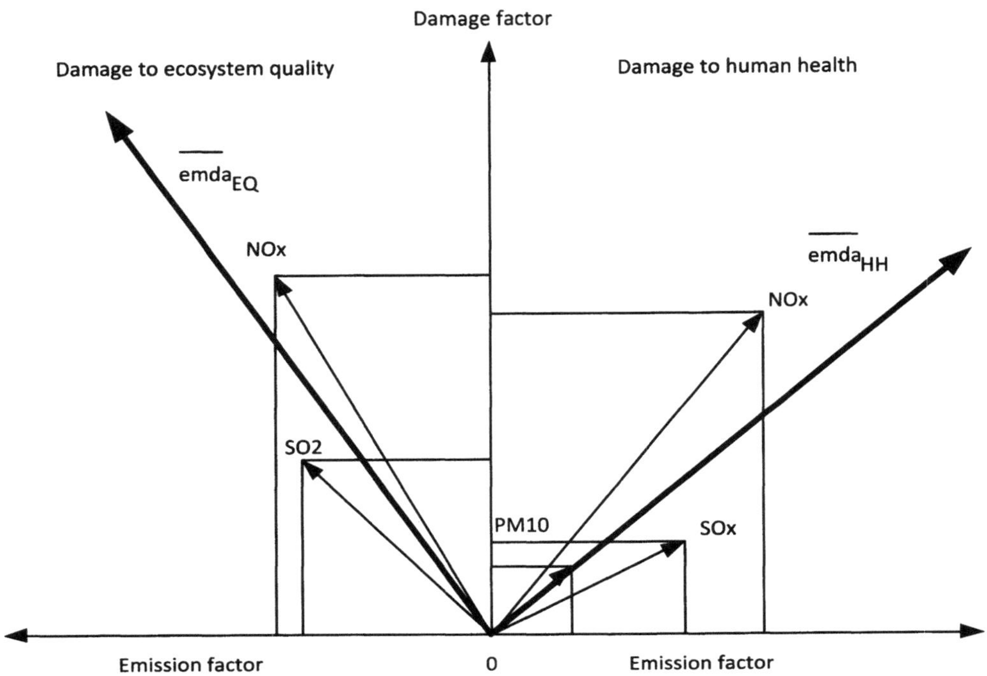

Integrating emissions and damage

In order to effect the *emda vector* approach, the authors have applied the Eco-indicator 99 method and used the EMEP/CORINAIR vehicular emissions database.

Eco-indicators 99 is a damage-oriented method for Life Cycle Impact Assessment (LCA) in production and consumption patterns in Europe. The group of experts has developed damage models for the most important environmental impact categories. The methodology report from 2001 includes an extensive list of substances' damage factors that can be found in most popular LCA databases. The expert panel has categorised and weighted the following three types of environmental damages: human health, ecosystem quality and resources.

Damage to human health was expressed as DALY (Disability Adjusted Life Years). The damage factor is expressed per kg of emissions.

Damage to ecosystems quality is expressed as the percentage of species that have disappeared in a certain area due to the environmental situation: PDF (Potentially Disappeared Fraction). The estimation of damage factor is based on kg of emissions.

Resource extraction is related to a parameter that indicates the quality of the remaining mineral and fossil resources. In both cases, the extraction of these resources will result in higher energy for future extraction.

The authors of this article have applied the Eco-indicator 99 method. Due to the high complexity of the issue, the authors have limited themselves to the local effects of ambient air pollution, i.e. estimated damage to human health and damage to ecosystems quality. The respective damage factors, based on a hierarchical model, are presented in Table 1.

Table 1: Damage to human health and ecosystem quality

	Human health (DALY)	**Ecosystem quality (PDF*m2*per year)**
SOx	5.46E-05	1.041
NOx	8.87E-05	5.713
PM10	3.75E-04	0

EMEP/CORINAIR emission factors were used to calculate emissions produced by the exhaust systems of road vehicles. Emission factors were worked out by the members of the CORINAIR Working Group.

Table 2: Vehicular emission factors typical for Kaunas

mg/km	**SOx**	**NOx**	**PM_{10}**
LDV gasoline	0.056	0.113	0
LDV diesel	0.19	0.58	0.043
LDV liquid petrol gas	0	0.113	0
Microbuses diesel	0.246	1.037	0.061
HDV diesel	0.314	1.301	0.115
Buses diesel	1.013	5.865	0.177

As a result of data analysis, based on the age of the motor vehicle fleet in Kaunas, the authors considered that average emission factors in Kaunas should correspond with Euro II 96/69/EC technology. Respective emission factors are presented in Table 2.

Analysis of traffic flows in Kaunas

The analysis of the traffic flows was based on the automated count of vehicles in 20 streets of Kaunas [4]. The stationary counter-classifier Marksman 660 (Golden River Traffic, UK) was used to identify 24-hour averages in the different vehicle classification groups. The study involved measurements of traffic intensity according to the EUR6 classification system. In order to effect this system, the authors have derived four classes of motorised vehicle: light duty vehicles (LDV), microbuses (passenger and goods), heavy duty vehicles (HDV) and buses (see Fig. 3).

Table 3. Intensity of traffic flows (yearly 24-hour averages)

	Type				
No	**LDV**	**Micro buses**	**HDV**	**Buses**	**Total**
1	30644	1728	571	0	32943
2	17425	2989	1843	0	22257
3	16444	6215	1629	90	24378
4	9724	800	578	30	11132
5	8640	1622	499	63	10824
6	20251	4478	687	152	25568
7	3903	1208	365	84	5560
8	21126	1925	557	608	24216
9	9919	2231	130	130	12410
10	4658	6428	191	382	11659
11	3882	867	534	24	5307
12	13750	2829	760	261	17600
13	20437	4554	1005	267	26263
14	16105	4432	1065	106	21708
15	5186	1261	180	14	6641
16	6704	3472	204	315	10695
17	12300	2084	207	251	14842
18	9072	3452	339	162	13025
19	16544	6415	3247	231	26437
20	12300	2084	207	251	14842

The data sets revealed that the seasonal traffic intensity for the city of Kaunas, based on 24-hour time series with the exceptional case of summer vacations, is rather stable.

The traffic intensity values for working days exceeded weekend values by 1.44-1.67. The minimal values of the traffic flows were recorded on Sundays and comprised 60.7 per cent of week averages.

According to the national statistics on the fuel types used, light duty vehicles (LDV) were brought into three categories: driven on gasoline – 50 per cent, driven on diesel fuel – 30 per cent, driven on liquid petrol gas – 20 per cent. It was considered that the rest motor vehicle classes, i.e. microbuses, heavy duty vehicles (HDV) and buses, were driven on diesel fuel.

Vehicle kilometres of travel

Analysis of cities across the globe illustrates a strong relationship between vehicle kilometres of travel (Lyons, T.J. et al., 2003) and urbanised land area. As VKT can be used as a surrogate for vehicular emissions, this provides a method for estimating urban vehicular pollution.

Cameron and his team (Cameron, I. et al., 2003) defined a non-dimensional private motorised mobility indicator for a city: the ratio of actual total annual private motorised passenger kilometres to the potential total annual private motorised passenger kilometres (see Box 3).

The term S is a vehicle saturation factor that changes relatively slowly as a function of the number of cars per head of population, but nevertheless represents an infrastructure limit on the growth of total vkt. It is a measure of automobile ownership and illustrates the impact of economic growth through higher vehicle ownership. Implicit within this model of private vkt is the impact of the saturation of car ownership as well as population growth.

Table 4: Π_{mob} values for car fleet in Kaunas

	Year				
Type	**2001**	**2002**	**2003**	**2004**	**2005**
LDV gasoline	0.192	0.205	0.222	0.235	0.249
LDV diesel	0.115	0.123	0.133	0.141	0.149
LDV liquid petrol gas	0.077	0.082	0.089	0.094	0.100
Microbuses diesel	0.097	0.103	0.112	0.119	0.126
HDV diesel	0.023	0.025	0.027	0.028	0.030
Buses diesel	0.007	0.007	0.008	0.008	0.009

The administrative area of the city of Kaunas makes up 157.15 km^2 and the number of inhabitants is approaching 400,000. The rapidly growing number of individual cars is identified as a major contributor to the ambient air pollution, especially in the central part of the city.

Calculations of Π_{mob} for private motorised mobility in Kaunas were combined with the data on traffic flows. As a result, Π_{mob} for private motorised mobility and other motor vehicle classes, i.e. microbuses, HDV and buses were estimated (see Table 4). It was assumed that proportions between numbers of motor vehicles in different classes would stay the same throughout the investigation period.

Results and discussion

The authors have derived emda vectors for human health and ecosystem quality. Emda absolute value for human health amounts to 14.30 DALY/km; for ecosystem quality it amounts to 41.95 PDF·m^2·year/km. It should be noted that the authors have composed emda from three contaminants (SOx, NOx and PM10). These contaminants are the principal ones when estimating respiratory effects on human health (inorganic compounds) and acidification-eutrophication effects on ecosystems; however if we extend our investigation and estimate carcinogenic effects on human health, or toxicity effects on ecosystems, the list of contaminants would comprise several dozens of chemical compounds.

Table 5: Motorized mobility indicators

	Indicator			
Year	**S**	$\Pi_{vehicle}$	Π_{uf}	Π_{mob}
2001	0.46	0.403	0.048	0.511
2002	0.49	0.428	0.047	0.545
2003	0.53	0.464	0.047	0.591
2004	0.56	0.493	0.046	0.625
2005	0.59	0.532	0.046	0.663

The indicators in the Table 5 present yearly changes in motorised mobility values in Kaunas during the period 2001-2005. The analysis of vehicle saturation factor (S), passenger vehicle kilometres per urban street network ($\Pi_{vehicle}$), and ratio between actual motorised kilometres to the potential motorised kilometres (Π_{mob}), have shown different rates of growth; whereas urban form ratio (Π_{uf}) indicated small decline.

Fig. 3: Dynamics of urban motorised mobility induced impacts

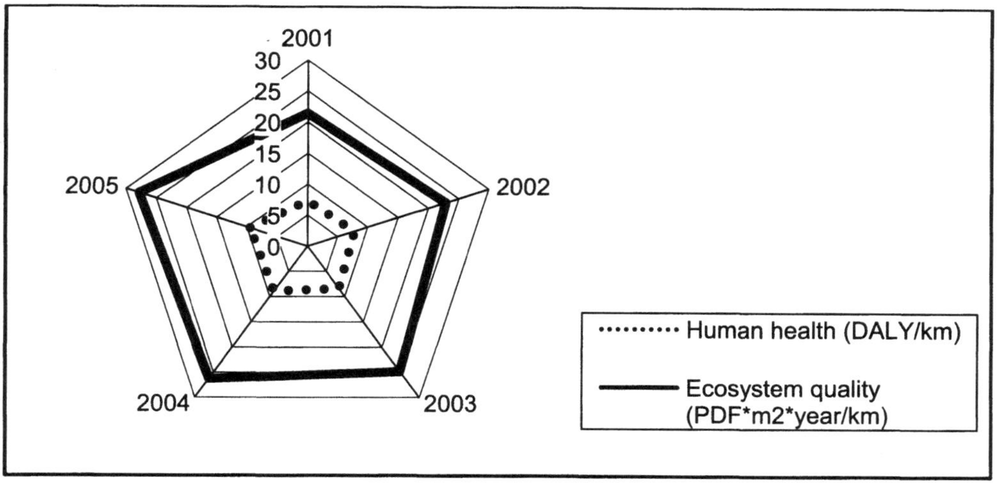

Emda vector approach was tested in Kaunas. Derived emda values were combined with the ratio between actual motorised kilometres to the potential motorised kilometres (Π_{mob}) for Kaunas. The spider diagram (see Fig. 3) illustrates the dynamics of urban motorised mobility induced impacts on respiratory effects on human health (inorganic compounds), and acidification-eutrophication effects on ecosystems. The analysis of urban motorised mobility induced impacts has showed constant rising of values per motor vehicle driven kilometre. It was also observed that the impact growth rate for the acidification-eutrophication effects is higher than that for human health.

Conclusions

The method of aggregation of vehicular emissions and damage factors in the form of vectors provides the new approach to the integrated assessment of environmental impacts.

The study has shown that derived *emda vector* could be applied to estimate the degree of damage of respiratory effects on human health (inorganic compounds), and acidification-eutrophication effects on ecosystems in an urban environment.

The calculation method of urban vehicle kilometres of travel (vkt), developed by Australian authors, was applied to Kaunas. The authors of this paper have combined vkt calculations with the statistical data on traffic flows; as a result, calculations of Π_{mob} were applied to different categories of motor vehicles.

References

Cameron, I., Kenworthy, J.R. and Lyons, T.J. (2003). Understandingand predicting private motorised urban mobility. *Transportation Research Part D* 8, 2003, pp. 267-283.

Commission's Guidance on the implementation of Directive 2001/42/EC on the assessment of the effects of certain plans and programmes on the environment (2003).

COST 350 action (2006). *Integrated Assessment of Environmental Impact of Traffic and Transport Infrastructure*. http://www.rws.nl/rws/dww/home/cost350.

Lyons, T.J., Kenworthy, J.R., Moy, C. and dos Santos, F. (2003). An international urban air pollution model for the transportation sector. *Transportation Research,* Part D 8, pp. 159-167.

Traffic intensity measurements using automatic counters-classifiers (Transporto priemonių eismo intensyvumo skaičiavimas automatiniais eismo intensyvumo skaitikliais-klasifikatoriais. Report. *Kaunas municipality* (2001), 93 p. (in Lithuanian).

VI. Current State and Future Visions

Urban traffic, especially private car use, has increased to a level where it is a main concern in city planning. Energy use, climate change, space intrusion, noise and urban air quality: these are the main environmental problems caused by urban transportation. These factors are also detrimental to human health and ecosystems. It is obvious that automotive fossil fuel dependent transport is not sustainable and has to undergo significant transformations (Kliucininkas, 2004).

Local environmental problems and effects of the remediation measures exhibit peculiarities specific to the country and region affected as well as to the time of occurrence. It is therefore valuable to know the experiences of other cities and to understand the underpinning explanatory mechanisms (Nakamura et al., 2004).

The solutions to sustainable urban transportation related environmental problems depend on the following preconditions:

- Understanding trends and features;
- Monitoring and modelling urban environmental quality;
- Recognising urban transportation management framework;
- Selecting appropriate tools for urban transportation management;
- Building interoperable systems between research & innovation and decision-making.

The monograph comprises the above-mentioned preconditions and presents an integrated approach to sustainable urban transportation planning and management. The materials presented address themselves not only to the enumeration of possible measures and tools, but also provide practical examples of research and development. The monograph follows the structure set out in Fig. 14.

Population growth and increasing incomes of urban inhabitants are expected to be the future drivers of urban transportation in 21st Century. Increasing demand for private mobility, particularly automobile transport, is likely to pose many challenges to long-term sustainable development. In particular, the need to reduce greenhouse gas emissions and to manage non-renewable energy resources are two challenges incompatible with a large expansion of automobile use (Ryan and Turton, 2007).

May and co-author (May and Matthews, 2001) distinguished between the way in which policy instruments for improved sustainability work, and identified four categories of policy instruments (each with a number of subcategories): technological, regulatory, informative and economic.

Fig. 14: Chapter structure

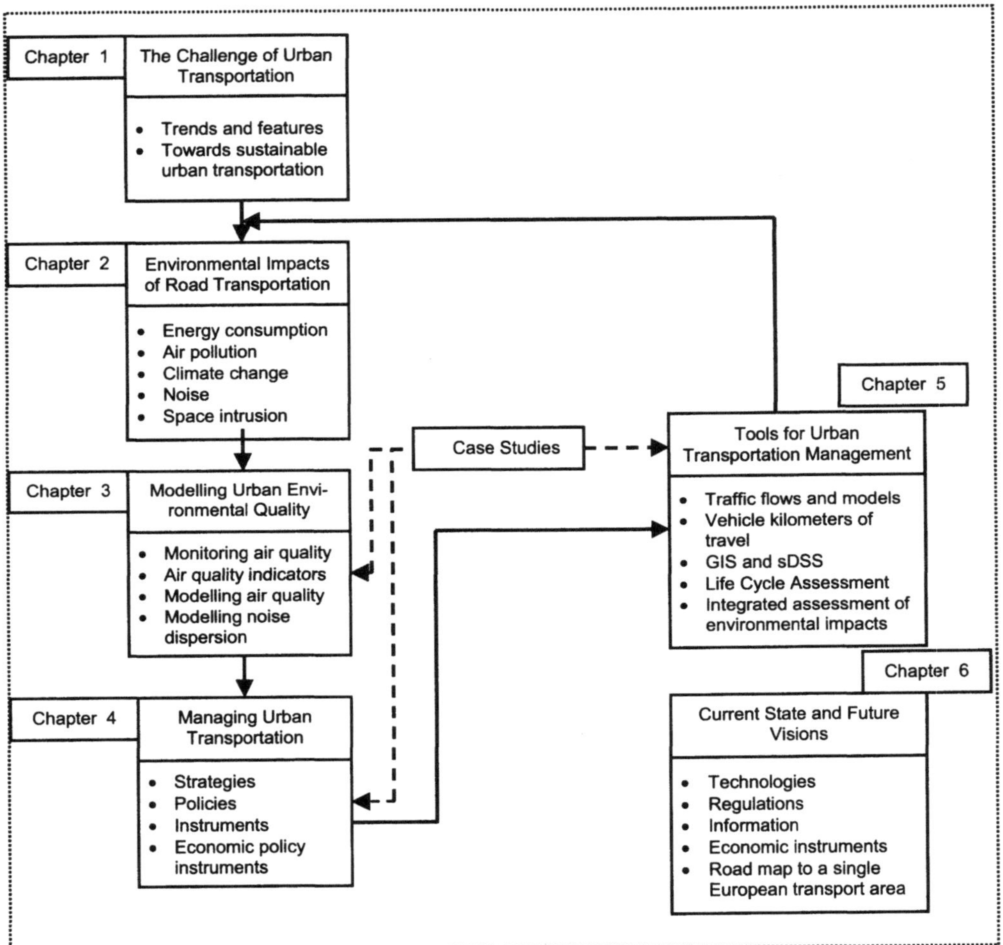

6.1 Technologies

The potential for far-reaching impact of technology is perhaps best appreciated by considering the role of the internal combustion engine (ICE) in the 20th century. The ICE was one of the most important technologies of the century related to transportation; moreover, the ICE also transformed global energy production. The ICE is a mature technology that has undergone a massive amount of technological developments, and continues to receive enormous R&D support, which is likely to sustain its dominance of the automobile and lorry for at least the medium term.

6.1.1 Internal combustion engine technology

Natural gas is a relatively well-tested alternative to petroleum in transport. A relatively small amount of natural gas is used in road transport. This fuel has many advantages: gas is clean-burning; existing distribution infrastructure is available in many regions, and it has great potential to reduce dependence on imported oil properties (although CNG vehicles will clearly increase their dependence on gas). Natural gas, particularly in regions where pipelines already exist, may provide a solution – in future perhaps combined with the separation and storing of carbon dioxide (sequestration) – until new technologies mature.

Alcohol fuels, typically methanol and ethanol, are also proven alternatives to petroleum-based fuels. In the long run, methanol has the potential to play a role in a possible transition to a hydrogen economy, where hydrogen replaces fossil fuels as the most important energy carrier, because it can be readily used as a feedstock for hydrogen production. The methanol industry is working on technologies that would allow methanol to be used as a source of hydrogen for fuel cell vehicle applications (AFDC, 2003).

Bioethanol is another alcohol fuel, similar to methanol although less toxic and corrosive. Blending bioethanol with gasoline increases the fuel octane rating and results in better oxygenation, thereby improving combustion and emission quality: in particular reducing ozone pollution (AFDC, 2003).

Hydrogen is one more ICE fuel option. This alternative is receiving attention because ICE hydrogen-fuelled vehicles produce no direct greenhouse emission, although they still produce some pollutants such as NO_x. Moreover, hydrogen can be produced from a variety of primary energy sources, reducing dependence on conventional transport fuels such as oil. *For these reasons there is much interest in the concept of a "hydrogen economy" in the future that will solve greenhouse gas and energy supply problems.*

6.1.2 Electric vehicle technology

The main attraction of electric vehicles is the complete absence of tail-pipe emissions, and consequently the promise of improved urban air quality. A further advantage is that using electricity to power vehicles affords much greater primary fuel flexibility, reducing dependence on petroleum products, but importantly also enabling a vehicle to be powered ultimately by renewable energy sources. Consequently, electric vehicles have the potential to be essentially pollution-free on both a well-to-tank and tank-to-wheels basis (IEA, 1997). Electric vehicles are also able to achieve very high levels of efficiency compared to ICE vehicles. This is because they avoid the large thermodynamic energy losses associated with the conversion of the chemical energy stored in fuel into mechanical power in ICEs. The critical issue for electric vehicles, however, is how to store and supply the electric power for the electric motor.

Battery electric vehicles store the electricity used to power the vehicle in on-board batteries, and these batteries are recharged by connecting the vehicle to the electric grid (AFDC, 2003). The potential impact of battery electric vehicles on long-term sustainability is therefore highly dependent on the fuels and technologies used in electricity production.

Fuel cell vehicles (FCVs) represent a radical departure from vehicles with conventional internal combustion engines. Like electric vehicles with batteries, FCVs are propelled by electric motors, and are also able to achieve high system efficiencies: up to 60 per cent, compared to around 20 per cent for petrol ICEs. But while battery electric vehicles use electricity from external sources, FCVs generate their own electric power on board through an electrochemical process using hydrogen fuel (pure hydrogen or hydrogen-rich fuel such as methanol, natural gas or even petrol) and oxygen from the air.

6.1.3 Hybrid vehicles

The fuels and technologies discussed so far fall into two largely separate groups: ICE and electricity-based vehicles. However, it is also possible to combine, or hybridise, mechanical and electrical conversion technologies on board a single vehicle to exploit the advantages of each.

6.2 Regulation

There are a number of policy instruments under this heading; we will concentrate on several essential ones. Most day-to-day urban mobility in any middle-sized city takes place within local-regional labour and service markets. The urban transport system can be discussed in terms of distance. Distance-wise urban transports, for both people and goods, may be subdivided into:

- sub-local (inside parts of the city) – 0-5 km,
- local (between parts of the city) – 5-40 km,
- regional (around the city) – 40-250 km.

Thus, mobility on this level should be reformed, the urban areas being organised in such a way makes it possible to reduce transport needs and make transport modes more sustainable. The city may therefore, in that sense, be planned in a zoning system, where the transport-oriented companies will get good access to the national and international transport systems, while others can be located in more densely built areas, not so favourable for car traffic. In other words, services are created closer to prime consumers. The city should be multifunctional in a way that many visit points are close to the final destination. This could be achieved by optimising road networks (expanding or reducing traffic capacity) and infrastructures.

An interesting question of urban morphology is: how much – and in what respect – may a real city diverge from this model and still include urban qualities?

Parking controls can be imposed in several ways. On-street parking controls contribute to improving the road network by increasing capacity and reducing car use; off-street parking controls are primarily concerned with reducing car use. A limit can be placed on the number of spaces, either in developments or by closing existing spaces. All of these measures can be combined with parking charges.

If it becomes common practice to let idle cell cars and other cell-fuelled vehicles generate electricity and heat while they are parked, there will be a need to arrange for cell vehicle parking in city blocks and at park & ride facilities. Vehicles will more than pay for their parking space; the distance between generator and the space to be heated will be very short, and the vehicles will be parked under safe and secure conditions.

Urban traffic controls should be used to give public transport priority in urban networks. In terms of strategy, their main impact is to improve alternative modes, but they are also often designed as ways of reducing car use. In managing public transport services, operators can use the same fleet to provide higher frequency on fewer routes, or lower frequency on a denser network. They are also able to influence the quality and reliability of their services. The significant improvements from the point of view of passengers should be in frequency of service (including small neighbourhoods outside the city proper) and in opportunities to travel directly, without the inconvenience of transfers.

6.3 Information

Information involves the provision of information to potential and actual users of the transport network to improve their use of the system or to encourage them to use it in different ways. The **K**nowledgebase **on** **S**ustainable **U**rban **L**and use and **T**ransport (konSULT) initiative identifies ten types of information provision, from conventional signposting and timetables to real-time information provision, and six types of "enlightenment" approach, which are designed to influence people to travel in different ways.

The information communication technologies (ICT) provide exciting opportunities for urban traffic planners and researchers (Limoges et al., 2000). The areas which face rapid development are:

- Intelligent transportation systems;
- Automated data collection;
- Spatial structures and mobility;
- Geographic Information Systems;
- Decision Support Systems and scenario building.

Real-time information allows drivers to make timely decisions. Surveys suggest that a well-designed information system may save up to 10 per cent of travel-time in congested areas for drivers who follow the guidance given to them.

It appears that teleworking provides a potential alternative to travel as a means of communication; therefore it could have significant environmental quality, safety and congestion reduction benefits. Experience suggests that as much as 35 per cent of car use could, in due course, be replaced by telecommuting.

6.4 Economic instruments

Economic instruments use pricing and taxation to influence use and to raise revenue for investment.

Taxes and charges applied to transport are well-established practices, which are well-studied in the literature. Market-based instruments can provide an efficient way to reduce emissions from transport by encouraging drivers to purchase fuel-efficient vehicles and to drive less. Fuel taxes and vehicle taxes affect drivers' travel behaviour and reduce vehicle kilometres of travel, whereas vehicle taxes tend to influence vehicle-purchasing decisions only. Distance-based and parking charges are effective, as they have an impact on vehicle kilometres of travel and hence reduce congestion, accidents, noise and emissions.

With very few exceptions, public transport users have always had to pay for the service which they receive, but the ways in which this is done vary considerably. The traditional distance-based fare structure has been replaced by flat fares and zonal fares, and payment per journey has been replaced by daily, monthly and yearly passes.

The vision offered here takes its point of departure in the relationship between economy of resources and quality of life. Quality of life that is sustainable may be termed "liveability": the community's ability to satisfy life's day-to-day needs. Liveability is a question of accommodating individuals' social and economic activity within the framework of such sustainability as finite resources will allow (Hulten and Karlstrom, 2004).

6.5 Roadmap to a Single European Transport Area

The European Commission adopted a roadmap of 40 concrete initiatives (White Paper, 2011) for the next decade to build a competitive transport system that will increase mobility, remove major barriers in key areas and fuel growth and employment. At the same time, the proposals will dramatically reduce Europe's dependence on imported oil and cut carbon emissions in transport by 60 per cent by 2050.

By 2050, key goals will include:

- No more conventionally fuelled cars in cities,
- 40 per cent use of sustainable low carbon fuels in aviation; at least 40 per cent cut in shipping emissions,
- A 50 per cent shift of medium distance intercity passenger and freight journeys from road to rail and waterborne transport,
- All of which will contribute to a 60 per cent cut in transport emissions by the middle of the century.

In cities, switching to cleaner transport is facilitated by the lower requirements for vehicle range and higher population density. Public transport choices are more widely available, as well as the option of walking and cycling. Cities suffer most from congestion, poor air quality and noise exposure. Urban transport is responsible for about a quarter of CO_2 emissions from transport, and 69 per cent of road accidents occur in cities. The gradual phasing out of "conventionally fuelled"* vehicles from the urban environment is a major contribution to the significant reduction of oil dependence, greenhouse gas emissions and local air and noise pollution. It will have to be complemented by the development of an appropriate fuelling/charging infrastructure for new vehicles.

A higher share of travel by collective transport, combined with minimum service obligations, will allow the density and frequency of service to be increased, thereby generating a virtuous circle for public transport modes. Demand management and land-use planning can lower traffic volumes. Facilitating walking and cycling should become an integral part of urban mobility and infrastructure design.

The use of smaller, lighter and more specialised road passenger vehicles must be encouraged. Large fleets of urban buses, taxis and delivery vans are particularly suitable for the introduction of alternative propulsion systems and fuels. These could make a substantial contribution in reducing the carbon intensity of urban transport while providing a test bed for new technologies and an opportunity for early market deployment. Road pricing and the removal of distortions in taxation can also assist in encouraging the use of public transport and the gradual introduction of alternative propulsion.

The interface between long-distance and last-mile freight transport should be organised more efficiently. The aim is to limit individual deliveries, the most "inefficient" part of the journey, to the shortest possible route. The use of Intelligent Transport Systems contributes to real-time traffic management, reducing delivery times and congestion for last-mile distribution. This could be performed with low-emission urban lorries. The use of electric, hydrogen and hybrid technologies would not only reduce air emissions, but also noise, allowing a greater portion of freight transport within the urban areas to take place at night time. This would ease the problem of road congestion during morning and afternoon peak hours.

References

AFDC (Alternatyive Fuels Data Center) (2003). *Alternative fuels, information and comparisons*. US Department of Energy, http://www.afdc.doe.gov/altfuels.html.

Hulten, P. and Karlstrom, M. (2004). New Technologies for Sustainable Urban Transport Systems. A vision of fuel cells as deliverers of improved urban quality. *Baltic University Urban Forum*, pp. 22-28.

IEA (International Energy Agency) (1997). *Energy technologies for the 21st Century*. Paris.

* The term "conventionally fuelled" refers to vehicles using non-hybrid, internal combustion engines (ICE).

Kliucininkas, L. (2004). Managing Urban Transportation – means to improve sustainability. *Baltic University Urban Forum,* pp. 29-34.
Limoges, E., Purvis, C.L., Turner, S., Wigan, M. and Wolf, J. (2000). *Future of Urban Transportation Data*. Millennium Papers of TRB A1D08: Committee on Urban Transportation Data and Information Systems.
May, A.D. and Matthews, B. (2001). *Initial Policy assessment*, PROSPECTS Deliverable 4, Leeds, Institute for Transport Studies.
Nakamura, H., Hayashi, Y. and May, A.D. (eds.) (2004). *Urban Transport and the Environment*. Elsevier, pp. 4-5.
Ryan, L. and Turton, H. (2007). Sustainable Automobile Transport. *Shaping Climate Change Policy*. Edward Elgar, pp. 38-53.
White Paper (2011). *Roadmap to a Single European Transport Area – Towards a competitive and resource efficient transport system*. COM (2011) 144 final.

Peter Lang · Internationaler Verlag der Wissenschaften

Martin Jungwirth

Environmental Management Systems in Local Public Authorities

A Comparative Study of the Introduction and Implementation of EMAS in the United Kingdom and Germany

Frankfurt am Main, Berlin, Bern, Bruxelles, New York, Oxford, Wien, 2011.
264 pp., num. fig., tab. and graph.
Contemporary Problems of Modern Societies.
Edited by Peter Nitschke and Corinna Onnen-Isemann. Vol. 5
ISBN 978-3-631-59531-2 · hb. € 44,80*

This study analyses the introduction and implementation processes of the European environmental management standard EMAS in modern local public administrations in the UK and Germany. Based on the different political and institutional settings, a qualitative analysis reveals the effects of the policy instrument. The study comes to the conclusion that the organizational processes are mostly dependent on the dominant actors within each organization. The orientation towards New Public Management can be beneficial for organizations working with EMAS while others can use it to start a modernization process. On a national level, EMAS is a classical case of an innovative, voluntary and flexible policy instrument with an imperfect national implementation that has dramatic effects on its use.

Content: Local public authorities in the UK and Germany · Modernization of the public sector · Environmental policy · European environmental management system · Policy implementation

*The e-price includes German tax rate. Prices are subject to change without notice

Frankfurt am Main · Berlin · Bern · Bruxelles · New York · Oxford · Wien
Distribution: Verlag Peter Lang AG
Moosstr. 1, CH-2542 Pieterlen
Telefax 0041 (0) 32 / 376 17 27
E-Mail info@peterlang.com

40 Years of Academic Publishing
Homepage http://www.peterlang.com

For the avoidance of doubt, input of the whole or any part of this publication into any generative artificial intelligence model, chatbot, program, software or technology is expressly prohibited without the prior written consent of the publisher.